Games on Symbian OS

A Handbook for Mobile Development

Games on Symbian OS

A Handbook for Mobile Development

Authors

Fadi Chehimi, Leon Clarke (Ideaworks3D), Michael Coffey, Paul Coulton, Twm Davies, Roland Geisler, Nigel Hietala, Sam Mason, Peter Lykke Nielsen, Aleks Garo Pamir and Jo Stichbury (Lead author and editor)

Contributors

Sam Cartwright (Mobile Developer Lab), Tim Closs (Ideaworks3D), John Holloway (ZingMagic), David MacQueen (Screen Digest), Adam Taylor (Ideaworks3D) and Steve Townsend (Great Ape Software)

Reviewed by

Michael Aubert, Jan Bonnevier, Sam Cartwright, Martin de Jode, Brian Evans, Toby Gray, Martin Hardman, John Imhofe, Mathew Inwood, Mark Jacobs, Erik Jacobson, Rob James, Elina Koivisto, Kazuhiro Konishi, Mal Minhas, Ben Morris, Matthew O'Donnell, Matt Plumtree, Lane Roberts, Jesus Ruiz, Hartti Suomela, Steve Townsend, Shawn Van Every and Sally Vedros

Head of Symbian Press

Freddie Gjertsen

Managing Editor

Satu McNabb

Copy Editor

Lisa Voisin

John Wiley & Sons, Ltd

Copyright © 2008   Symbian Software Ltd

Published by   John Wiley & Sons Ltd, The Atrium, Southern Gate, Chichester,
West Sussex PO19 8SQ, England

Telephone (+44) 1243 779777

Other Wiley Editorial Offices

John Wiley & Sons Inc., 111 River Street, Hoboken, NJ 07030, USA

Jossey-Bass, 989 Market Street, San Francisco, CA 94103-1741, USA

Wiley-VCH Verlag GmbH, Boschstr. 12, D-69469 Weinheim, Germany

John Wiley & Sons Australia Ltd, 42 McDougall Street, Milton, Queensland 4064, Australia

John Wiley & Sons (Asia) Pte Ltd, 2 Clementi Loop #02-01, Jin Xing Distripark, Singapore 129809

John Wiley & Sons Canada Ltd, 6045 Freemont Blvd, Mississauga, Ontario, L5R 4J3, Canada

Wiley also publishes its books in a variety of electronic formats. Some content that appears
in print may not be available in electronic books.

British Library Cataloguing in Publication Data

A catalogue record for this book is available from the British Library

ISBN: 978-0-470-99804-5

Typeset in 10/12pt Optima by Laserwords Private Limited, Chennai, India
Printed and bound in Great Britain by Bell & Bain, Glasgow
This book is printed on acid-free paper responsibly manufactured from sustainable forestry
in which at least two trees are planted for each one used for paper production.

Contents

Part Two: Creating Native Games on Symbian OS v9

Part Three: Porting Games to Symbian OS

Part Four: Java ME, DoJa and Flash Lite on Symbian OS

Foreword

Tim Closs, Chief Technology Officer, Ideaworks3D Ltd

The mobile phone is not primarily a platform for playing games, and Symbian OS is not primarily a platform for writing games. So why are you here, at the opening pages of a book devoted to the subject of mobile games on Symbian OS?

I don't have to tell you about the revolution that has happened (and is still happening) in mobile phone hardware. Phones today are unrecognizable from the cumbersome black boxes we see in 1980's video clips, proudly displayed as fashion accessories by loud-mouthed stockbrokers. One mobile phone manufacturer today describes their flagship range as 'multimedia computers' – and we can see why. High resolution, sharp and brightly-lit screens, many megabytes of memory, fast processors, and even hardware graphics acceleration, these all add up to a very powerful piece of computing hardware. Whilst it may be true that voice communication and text messaging are still the most popular services, phones are increasingly being relied upon for email, web browsing, and entertainment such as music, video, mobile TV, and games.

Network operators derive revenue by encouraging users to download content to their phone. Until recently, the primary drivers for these revenues were 'wallpapers' (images displayed on the phone's front screen) and ringtones. However, as this book will explain, over the last year many operators have seen game revenues overtake other content types. This is due partly to the increase in the quality and variety of games on offer, but also to the improved user experience offered by operators to consumers when searching for and choosing to download a game.

Symbian OS is the leading smartphone operating system in the world today, occupying over 90 % of the European market, large swathes of the

Asian markets and elsewhere. In addition, Symbian OS has evolved over many, many years and is arguably a more mature and well-understood platform than many of its competitors. As such, Symbian OS is an excellent choice for mobile games developers looking to develop expertise and deploy to a wide consumer base.

Ideaworks3D is a leading developer of advanced mobile games and enabling technologies for cross-platform mobile game development. Our BAFTA award-winning studio collaborates with the industry's leading mobile and video game publishers to mobilize their flagship game franchises, including **Final Fantasy VII** (Square Enix), **Need for Speed**™ and **The Sims 2**™ **Mobile** (EA Mobile). We also innovate to create original games for handheld and mobile platforms, including such games as **System Rush**™: **Evolution** (Nokia). We have been developing games for Symbian OS since 2002. We were heavily involved in the launch of Nokia's N-Gage game decks (a handset designed specifically for playing mobile games, based on Symbian OS 6.1) in 2003. We are now equally involved in the next generation of N-Gage, as a platform that allows high-quality games on many of Nokia's flagship smartphones. We have also deployed games to all of the open native operating systems in the world today. As such, we believe we have some valuable insights into mobile game development on all platforms, including Symbian OS. We have passed on some of our experience within this book.

Our excitement for mobile games remains undimmed, and we believe the future for the industry is brighter than ever.

Foreword

Antony Edwards, VP Developer Product Marketing, Symbian Ltd

Game development has always been at the bleeding edge of technology. In 1984, when IBM first tried to enter the personal computer market with the IBM PCjr, they ensured Sierra On-Line's **King's Quest** was there for the launch. Game consoles were the first multi-processor computers to make it into most people's lives, and game developers have used the connectivity provided by the Internet to revolutionize their games more than any other software genre. We've seen **NetHack** on more prototype operating systems than we can remember. And, though it's not talked about, I'm sure that there was a game hidden away in the accumulators of the ENIAC somewhere. . .

Mobile devices present the next great adventure for game developers. Mobile devices are always connected, always at hand, and often include features such as location-based services and a camera, which provide a canvas for game developers to create the most engaging experiences yet.

This book is dedicated to helping game developers create and define this new genre of mobile games on Symbian smartphones. Symbian OS is the world's most popular smartphone operating system, having shipped in over 145 million devices across 120 models. Symbian licenses Symbian OS to the world's leading handset manufacturers and works closely with all leading companies across the mobile industry to help create new and compelling mobile experiences.

Symbian has been an innovator in multimedia and graphics since the beginning. Symbian OS was the first smartphone OS to support OpenGL ES for mobile 3D graphics in 2004. We recently announced a major new graphics architecture that supports OpenVG, OpenGL ES 2.0, and OpenWF Composition API. Symbian OS v9.5 is set to deliver higher performance and richer graphics. What will developers do with all our new technology? We're not entirely sure, but we *are* sure it's going to be exciting. That's the great part of being the world's leading mobile operating system.

Symbian is also making it easier to port existing games to our smartphone platforms. The P.I.P.S. and Open C initiatives for POSIX-compliant C/C++ development, support for standards such as OpenGL ES, and our open platform for middleware solutions all make Symbian OS flexible for professional developers to migrate games to our smartphones. What better demonstration of this could there be than Olli Hinkka's open source port of **Quake** for S60 3rd Edition Feature Pack 1 smartphones, which uses the P.I.P.S. libraries to recreate that pre-eminent shoot 'em up in all its gory glory?

The authors bring a wealth of experience from both the game industry and Symbian OS development to this book. Whether you are new to writing mobile games, or are already an experienced game developer, you will find it invaluable. It covers Symbian OS game development using C, C++, Java ME, DoJa, and Flash Lite. We hope it will spark your imagination to create games for our next generation of smartphones.

About this Book

This book forms part of the Symbian Press Technology series. It describes the key aspects of the mobile game market, with particular emphasis on creating games for smartphones based on Symbian OS v9.x.

What Is Covered?

This book divides into four parts. The first part introduces the world of mobile games. It aims to explain what mobile games are, who plays them, who writes and sells them, how they sell them and what the major issues are in the marketplace. We'll look at some statistics for the sales of mobile games and mobile phones, and make comparisons with game consoles, handheld systems and PC games. The first chapter is not technical and is suitable for anyone interested in finding out more about the mobile games industry. However, there is an excursion into the differences between BREW, Java ME and native C++ games, a brief foray into issues of compatibility and portability, and a short introduction to some of the aspects of game development that characterize any mobile platform. So there's something for developers too: topics are introduced gradually, with signposts directing readers where to go for more information within the rest of the book – and beyond.

The second part of the book covers various technical areas associated with creating games in C++ on Symbian OS v9 smartphones (using the native APIs provided by UIQ 3 or S60 3rd Edition SDKs). Chapter 2 covers the basics of writing a game in Symbian C++, Chapter 3 delves deep into the Symbian OS graphics architecture, Chapter 4 deals with adding audio to games, and Chapter 5 discusses the issues associated with creating a multiplayer game. The final chapter in this part of the book is Chapter 6, which discusses how to be innovative and create novel and appealing gameplay by using phone functionality, such as motion detection, the camera, the vibra or location-based services.

The third part of the book is for game developers who are interested in porting games without using just the native Symbian OS C++ APIs. Chapter 7 discusses the various standards support available on Symbian OS, such as POSIX-compliant standard C libraries, OpenKODE and OpenGL ES. Chapter 8 describes the Nokia N-Gage platform, which is Nokia's initiative to bring high quality games to S60 3rd Edition smartphones by providing a platform for professional game developers to port their game code using standard C and C++. N-Gage is more than an SDK and comprises an end-to-end solution for users to discover, play, and share games, and in Chapter 8, we'll discuss how developers work with Nokia to achieve this.

The final part of the book is for developers who want to write games for Symbian smartphones without using C or C++ at all. Chapter 9 describes the support available for Java ME on Symbian OS, and walks through an example game. The same author then explores the DoJa standards available for creating Java games for installation to Symbian smartphones in Japan, where the majority of phones are supplied by NTT DoCoMo and a different set of rules for application creation and distribution apply. The last chapter of this section, and the book, describes game creation using Flash Lite 2, which is supported on S60 3rd Edition smartphones and Symbian smartphones in Japan.

This book doesn't present a single example of a game that it builds from scratch throughout the book, because we find that this approach tends to constrain the text, and the reader, to the details of the example. Instead, we've used a number of different examples for each chapter; these have been tailored specifically to illustrate the topic in question. Where possible, we have avoided using large chunks of example code in the technical chapters of the book, and have instead put the code, in full, on the website for this book *developer.symbian.com/gamesbook*.

If you would like to read more about the creation of a full game example in C++ on Symbian OS, we highly recommend a paper on the Symbian Developer Network by one of the authors of this book, Twm Davies. The **Roids** paper (*developer.symbian.com/roidsgame*) explains the design, implementation and optimization of an **Asteroids** clone for Symbian OS v9. The example code and installation files for both S60 3rd Edition and UIQ 3 phones can also be downloaded from the website. Recommendations for other papers and code downloads for full game examples can be found in the References and Resources section at the end of this book.

Please also take a look at the book's page on the Symbian Developer wiki (***developer.symbian.com/wiki/display/academy/Games+on+Symbian+OS***) for a set of useful links to other mobile game developer resources, and an errata page for the book. Do feel free to visit it regularly and to contribute.

Who Is this Book for?

The typical reader may be:

- a C++ or Java ME developer already creating applications or services on Symbian OS who wants to take advantage of the growth in commercial mobile games

- anyone in the game industry (e.g., a professional or hobbyist developer, game producer or designer) who wants to target games for Symbian OS

- a developer new to Symbian OS who wants to learn about the platform and is experimenting by creating a game.

But we don't like to stereotype our readers, and hope that if you don't fit into these categories, you'll still find something of interest in this book!

The technical chapters assume that you have a working knowledge of either C++ on Symbian OS or Java ME. The basic idioms of Symbian C++ and details, such as how to get a working development environment or how to create 'Hello World,' are not to be found in this book, which is instead dedicated to the specifics of mobile game creation on Symbian smartphones. However, if you need general information about developing on Symbian OS, it can be found in other titles in the Symbian Press series, and in a number of free papers available on the Symbian Developer Network website and other Symbian OS community development websites. A list of the current Symbian Press series of books can be found in the References and Resources section at the end of the book.

About the Authors

Jo Stichbury

Jo has worked within the Symbian ecosystem since 1997, in the Base, Connectivity, and Security teams of Symbian, as well as for Advansys, Sony Ericsson and Nokia. At Nokia, she worked in the N-Gage team, providing technical support to game developers worldwide, and while there, she discovered that it is possible to play games at work legitimately. There's been no going back, and she's now trying to think up an excuse for playing **Spore** professionally for Symbian Press, whilst eagerly awaiting the game's release.

Jo is author of *Symbian OS Explained: Effective C++ Programming for Smartphones,* which was published by Symbian Press in 2004. She co-authored *The Accredited Symbian Developer Primer: Fundamentals of Symbian OS,* with her partner, Mark Jacobs, published by Symbian Press in 2006.

Jo became an Accredited Symbian Developer in 2005 and a Forum Nokia Champion in 2006 and 2007.

Twm Davies

Twm graduated in 1999 from Cardiff University with a First Class BSc in Computer Science, where he specialized in computer graphics and artificial intelligence. After uni, he moved to London where he worked at Symbian for seven and a half years as an engineer, consultant, and product manager helping Symbian licensees get their first phones out of the door.

At present, Twm is a freelance consultant on smartphone projects (or Ronin as he prefers to call it) and a regular contributor of book chapters and technical papers to the Symbian community. Twm wrote his first game at age 12 in Amos Basic, and his favorite game is **Chaos Engine** by the bitmap brothers (for its steam punk stylings).

Aleks Garo Pamir

Aleks graduated from Bogazici University in Turkey with an Associate degree in Computer Programming. He also has a BA in Labor Economics from Istanbul University and an MA in Industrial Relations from Marmara University. He spent a few years in Bogazici University as a teaching assistant in the MIS department.

During his education Aleks worked as a columnist and editor for local games review magazines in Turkey including PC World/Turkey. He was also a member of the team that developed the first Turkish RPG **Istanbul Efsaneleri - Lale Savascilari** for PCs. He worked as a software developer in Turkey for a number of years, developing software for a diverse set of industries using multiple different languages. After the emergence of cellular networks, he decided to focus on mobile technologies, and Symbian became his first choice.

Aleks moved to Canada in 2003 and founded Capybara Games, a mobile game company. Aleks became an Accredited Symbian Developer in 2007. He lives in Vancouver with his wife Zeynep and is currently working for Intrinsyc Software International as a senior software developer.

Michael Coffey

Michael joined Symbian as a graduate in 2004 after obtaining an MEng in Electronic and Electrical Engineering from the University of Birmingham. He has since worked in the PIM team specializing in the calendar and alarm server components. In his spare time, he has worked on creating Symbian mobile games. He is an Accredited Symbian Developer.

Leon Clarke

Leon Clarke has been working in embedded and mobile computing for over 15 years. He worked at Symbian for six years in various capacities, working on web browsing technologies and system software, before moving to Ideaworks3D four years ago, where he has been the chief architect of Ideaworks3D's online gaming product, Airplay Online. Leon has been actively involved in the Khronos Group's Open KODE standardization process, being one of the major contributors to the Open KODE core.

Paul Coulton

Paul is a senior lecturer based within the Department of Communication Systems at Lancaster University. When he first left university, Paul worked for various small games developer teams primarily on algorithm design, before utilizing his skills in the defence industry on simulators. In 1997, he completed a PhD in Mobile Systems and, although his early work was primarily associated with HDSPA, he switched to application development in 2000. The main focus of his current research surrounds innovative mobile social software with a particular emphasis on mobile entertainment, such as games. He was the first academic invited to speak at the Mobile section of the Game Developers Conference and was one of the founding Forum Nokia Champions in 2006, re-selected in 2007.

Paul has published widely (including a Symbian Press book on S60 development) and a lot of his research projects encompass novel uses of the latest technologies, such as RFID/NFC, cameras, GPS, and 3D accelerometers in mobile phones.

Because his research encompasses a great deal of HCI, Paul is a big fan of the Wii and the DS, although his favorite game is the old mega drive classic *Toejam and Earl2*, principally because it had a soundtrack by the master of funk, George Clinton.

Fadi Chehimi

Fadi is a mobile phone software engineer at Mobica Ltd. and a final year PhD student at the Department of Communication Systems at Lancaster University, UK. During his employment, Fadi has worked intensively on Symbian OS and Windows Mobile platforms. He has had his hands on several new devices and technologies before they were released into the market and for him this is part of the joy of mobile development. During his research he worked on several projects related to 3D graphics and mobile advertising. His main focus is on utilizing mixed reality technology for business and entertainment applications on mobile phones, and for that he has developed several proof-of-concept prototypes.

Sam Mason

Sam came to computing late in life, after spending three years studying civil engineering – where he first learned to program using Pascal. Unusually, he didn't have a Commodore 64, had never heard of a Z80, didn't write compilers at age 7, and still can't use a soldering iron! After an abortive attempt at trying to become a vet, Sam eventually graduated with a Computer Science degree from the University of New South Wales in Sydney in 1996.

He ran straight into the dot com boom as a contract programmer in eCommerce, building HR and legal extranets, as well as workflow and content management systems for a number of short-lived companies, using Java and C++. After spending about two years working on a Java-based multi-lingual video-on-demand system for Singtel and other APEC telcos, he's spent most of the last four years working with and learning about mobile phone technologies, while picking up a couple of Java certificates and completing a Masters of Information Technology in Autonomous Systems from UNSW along the way.

Having spent most of his professional career in fascinating sectors like accounting, finance, and payroll, he's become somewhat of a mobile technology and artificial intelligence evangelist these days. To that end, he started Mobile Intelligence in 2006 and became the first to sit the supervised Accredited Symbian Developer (ASD) exam in Australia, in March 2007.

Sam is happily married with three little ones who are all at pre-school, they have no cats or birds, even fewer dogs, and is currently working full-time as a mobile technology consultant based in Sydney.

Roland Geisler

Roland Geisler currently serves as Group Product Manager for Nokia and is responsible for the Nokia N-Gage product strategy and the global product management for the N-Gage application for Nokia smartphones. Roland joined Nokia in 2001, and has held various management positions in Europe and the US, including Head of Marketing and Strategy, and Technology Manager for the Nokia Mini Map Browser, project manager for a number of mobile application software projects, and Development Partner Manager for Opera Software, a Nokia software supplier. Before he joined Nokia, Roland worked as a product manager and software engineer at Gigabeat (acquired by Napster) in Silicon Valley. Prior to this, he was a research assistant at the National Center for Supercomputing Applications (NCSA). Roland earned an MS in Computer Science from the University of Illinois at Urbana-Champaign and his undergraduate degree in Computer Science and Economics from the Technische Universität München in Germany.

Peter Lykke Nielsen

Peter Lykke Nielsen started his career in the interactive entertainment industry back in early 1995, when he was part of the trio setting up and running the independent game developer Scavenger. As well as being part of building the company, he also oversaw a number of titles in development. In the late nineties, after having relocated to England, he joined Activision as a Producer and worked with them on **Rome: Total War** and **Rally Fusion: Race of Champions** plus a number of their Disney and Star Trek branded titles. In 2004, he relocated to Canada, where he initially worked as a Producer for EA on **Need for Speed: Most Wanted**. Subsequently, he joined Nokia to become part of the team that will launch the next generation of the N-Gage platform. In his current role as Product Manager for the N-Gage SDK, Peter Nielsen is responsible for researching, defining, and communicating the feature set required to create cutting edge N-Gage titles.

Samuel (Sam) Cartwright

Samuel graduated from Griffith University, Australia, with a BIT in 2000. Shortly thereafter, he joined an outsourcing firm in Tokyo specializing in development for the telecommunications industry. While there, he worked as a Windows application developer creating low-level protocol encoders and UML tools before a briefly working on a mobile phone MMI (man machine interface) using the Apoxi framework.

In 2005, Sam joined Gameloft K.K., where he is now a senior programmer. As a game programmer, Sam both codes original titles on MIDP and ports games from MIDP to DoJa. In his spare time he also runs the Mobile Developer Lab mobile programming site, and is completing an MBA from Charles Sturt University.

Outside of work, Sam enjoys working out at the gym and attempting to read manga in Japanese.

Nigel Hietala

Nigel Hietala is a User Interface Specialist working for Nokia on the S60 platform. He has worked for over a decade with mobile devices and always at companies using Symbian OS. He caught the programming bug while originally building UI prototypes of future designs with Flash MX and, much to the dismay of his designer colleagues, has been delving ever deeper into the world of Flash and software development.

He still loves computer games and imagines a time when his four children have grown up enough that he can play them again.

Editor's Acknowledgements

This book wouldn't have come into being without a great team to write it. We somehow managed to coordinate working in multiple time zones and countries (Australia, Canada, Finland, the UK and the US), and I'd like to thank Twm, Aleks, Michael, Leon, Paul, Fadi, Sam, Peter, Roland and Nigel for putting in countless hours on their contributions. We're all grateful to our families for their patience and understanding while we took time out to write this book.

We would all like to thank our reviewers for their efforts too, sometimes helping out at very short notice, and always providing insightful comments and showing great attention to detail. We'd particularly like to thank Sam Cartwright, not only for his review of the chapter about games in Japan, but also for volunteering to go to the 2007 Tokyo Game Show on our behalf, and writing a section about it.

We'd also like to thank our other contributors and collaborators, including David MacQueen of Screen Digest, Adam Taylor and Tim Closs of Ideaworks3D, Krystal Sammis, Jan Bonnevier, Jesus Ruiz and Carlos Hernadez-Fisher of N-Gage, and John Holloway of ZingMagic. Many other people took the time to talk to us about this project; thank you for your help – Erik, Steve, Phil, Simon, Neil, Bill and Tony at Symbian; Jonathan, Jeff, Van, Kevin and AaPee at Nokia. Thanks also to Annabel Cooke at Symbian for supplying some of the diagrams used in Chapter 1.

When the writing stops, the editing starts, and we owe a big 'thank you' to Lisa Voisin for her attention to detail and her serenity in the storm of our copy edit schedule. (Thanks must also go to Tierney at Nokia for persuading Lisa to work with us). We'd particularly like to acknowledge the hard work put in by the Symbian Press team: the ever-patient Satu McNabb and ever-calm Freddie Gjertsen. Thanks also to Mark Shackman and Rodney DeGale for their assistance with all things related to the Symbian Developer Network website.

Many thanks to Shena Deuchars for proofreading this book and to Terry Halliday for creating the index. Finally, a big 'thank you' to our counterparts at Wiley, particularly Rosie Kemp, Drew Kennerly, Colleen Goldring, Sally Tickner and Hannah Clement, for keeping us on track. May our page count estimates always be accurate...

Part One

**A Symbian Perspective
on Mobile Games**

1

Introduction

Jo Stichbury

1.1 Why Games?

Electronic games, or video games as they are sometimes known, are big business. The sales of games for PCs, consoles, portable game players and mobile phones are now competing with the film industry for consumer spending on entertainment per year. In September 2007, Microsoft's *Halo 3* became the fastest-selling computer game, generating global sales of $300 million worldwide in the first week it was released. The game was released for Microsoft's Xbox 360 console on September 25, and it generated global sales of more than $170 million in its first 24 hours.[1]

Let's compare the sales of games with movie tickets. In 2006, the US box office reported total sales to be $9.49 billion.[2] Over the same period, again in the US alone, the NPD Group reports game sales (portable, console and PC games) to be $7.4 billion (and sales of all games plus portable and console hardware, software and accessories, to have generated revenues of close to $13.5 billion).[3] And it's not just North America; in Japan, average monthly leisure spending was estimated to be 7300 yen on mobile phone fees, 3700 yen on an Internet connection, 2300 yen on electronic games, 2200 yen for books and other print media, 1600 yen for music, 1300 yen on karaoke and just under 1000 yen on movies.[4]

There is crossover and symbiosis between the industries. A good example comes from the animated motion picture *Cars*, made jointly by

---

[1] *www.microsoft.com/presspass/press/2007/oct07/10-04Halo3FirstWeekPR.mspx*
[2] *mpaa.org/researchStatistics.asp*
[3] *www.npd.com*
[4] *The Business and Culture of Digital Games: Gamework and Gameplay*, Aphra Kerr, Sage Publications Ltd, 2006

Disney and Pixar (now a subsidiary of Disney). The film made sales of $244.1 million in the US, the second highest figure of any movie released in 2006. A video game based on *Cars*, published by THQ, achieved the second highest console title sales in the US in 2006, and shipped more than seven million units.

Given the size of the revenue earned, it's interesting to think that the game industry is still quite immature compared to the film industry. There are some arguments over what counts as the first computer game ever created, but it's commonly held that the first digital video game dates back to 1961. **Spacewar** was written by Steve Russell at Massachusetts Institute of Technology, and ran on an early DEC minicomputer, which was not really intended for playing games! The first popular home console system was the Atari 2600, released in 1977.

That the game industry has caught up with the movie industry in such a short amount of time probably reflects the growing consumer trend for interactive entertainment. It is no longer sufficient simply to consume entertainment; people want to join in too. Games are a perfect medium to do this; they provide solo and multiplayer interaction over short periods of time, or for more prolonged sessions that can be returned to, as an 'alternative reality.'

Games are everywhere. They are played for entertainment on game consoles in the living room, or on handheld systems and mobile phones on the move. Games are used in schools or in the workplace as an educational or training aid. They can be played for a short duration, to pass time, or for hours at a time, as a player developers a character and skills in a long term role-playing or adventure game. Games are fun, but the game industry is a serious billion-dollar business.

1.2   What Is Symbian? What Is Symbian OS?

Symbian is a British company, formed in 1998 as a collaboration between Nokia, Ericsson, Motorola and Psion. The company supplies Symbian OS, which is the leading open operating system found in advanced data-enabled mobile phones, known as smartphones. At the time of writing this book, Symbian OS has been used in over 120 different models of smartphone.

Symbian does not make smartphones itself, but licenses Symbian OS to the world's major handset manufacturers (in alphabetical order, Fujitsu, LG Electronics, Mitsubishi Electric, Motorola, Nokia, Samsung, Sharp and Sony Ericsson).

Symbian OS is found in the majority of smartphones available. Gartner estimated Symbian's market share as 70 % in 2006, while Canalys put it slightly higher, at 71.7 % globally for the same period and, in a separate report, at 72 % in 2007. Regional figures for the smartphone market share in China were reported at 61 % for Q2, 2007, also by Canalys, and at

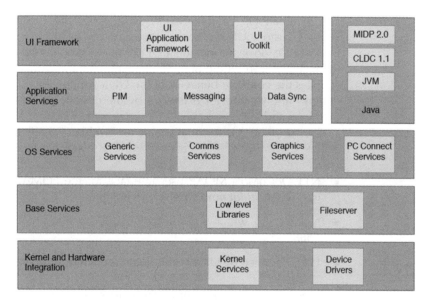

Figure 1.1   Representing Symbian OS using a layered model

72 % in Japan (where 20 million 3G Symbian smartphones have sold since 2003).[5]

Symbian OS is a modular operating system, which means that it is constructed from well-defined, discrete parts, which combine to allow for configurability across a range of target hardware. At a high level, Symbian OS can be thought of as a layered model, as shown in Figure 1.1, with hardware layers at the bottom and support for user-level applications at the top. From the bottom up, it comprises the following:

- kernel services and hardware interface layer

- base services such as file server, database and utilities libraries

- services for communication, graphics, multimedia, connectivity and standards support

- application services for personal information management (PIM), networking, messaging and application-level Internet protocols and other standards support

- user interface (UI) framework layer.

Symbian OS has a flexible architecture that allows different user interface layers to run as platforms on top of the core operating system.

---

[5] Figures were taken from Symbian's first and second quarter results for the year 2007, posted on their website at **www.symbian.com/about/fastfacts/fastfacts.html**

The Symbian OS UI framework just supplies the common core that enables custom UIs to be developed on top of the operating system. Symbian OS does not include a UI *per se* – it supplies the ability for one to be customized on top of it. The generic UI framework of Symbian OS supplies the common behavior of the UI: a windowing model, common controls and support for extension by the custom UI, which defines its own look and feel.

The custom UI platforms available for Symbian OS are, in alphabetical order:

- NTT DoCoMo's MOAP user interface for the FOMA™ 3G network (see *www.nttdocomo.com*) found in Japan

- Nokia's S60, which was formerly known as Series 60 (see *www.s60. com*)

- UIQ, designed by UIQ Technology (see *www.uiq.com*).

Symbian OS development environments range from the Symbian OS Customisation Kit (CustKit) which is provided by Symbian to handset manufacturers for them to create phone products, to SDKs which are provided by the UI platform vendors to application developers, and are used to create third party software for users to install. The latter can be obtained free of charge from the links found at *developer.symbian.com/main/tools/sdks*.

1.2.1 What Is a Smartphone?

Having briefly described Symbian, the company, and Symbian OS, the heart of a smartphone, let's talk more about what a Symbian smartphone actually is. This book discusses how to write mobile games specifically to take advantage of the capabilities of Symbian smartphones – but what does that mean in practice?

Symbian defines a smartphone as:

> 'a mobile phone that uses an operating system based on industry standards, designed for the requirements of advanced mobile telephony communication on 2.5G networks or above.'

The combination of such a powerful software platform with mobile telephony enables the introduction of advanced data services and innovation by device creators. It also allows the user to personalize the phone by installation of a wide range of applications, which in turn opens up the market for an entire industry of mobile software developers (such as game developers). A user can transform a smartphone into a unique and personal tool to suit their requirements. It may become a business productivity tool (with Push email, an office suite and even applications written by them to be specific to their business) in one person's hands. The same phone may alternatively be a mobile entertainment device (for games, videos, mobile TV or music) or an in-car navigation system, depending on the user's lifestyle.

Nokia often refers to their Nseries S60 devices as 'Multimedia Computers' because they offer the functionality of a PC combined with those supplied by many portable single-purpose devices. For example, Nseries devices (and other Symbian smartphones) may include features such as WLAN, email, high-quality (5 megapixel) cameras, video capture and playback, mobile TV and music players with storage of up to 8 GB. Other functionality offered by Symbian smartphones may include GPS, visual radio, Bluetooth and USB local connectivity, accelerometer, Java, Web and WAP, MMS, SMS and, of course, telephony (voice and 3G data). Symbian smartphones come with a range of embedded applications such as calendar, address book, photo viewers (and editors), music players, messaging (SMS, MMS, email), web browser, converters, sound recorder and many others. Most also have at least one built-in game.

Symbian smartphones are frequently known as 'convergence devices' because the smartphone can take on the role of other single-purpose devices, rather like a Swiss army knife, and it removes the need to carry around more than one gadget. It may seem amazing, but the largest digital camera manufacturer in the world is Nokia[6] because of the number of camera phones it sells annually. Besides cameras, smartphones can now take the role of alarm clocks, calculators, pagers, game consoles, music players, portable radios, video cameras, pedometers, dictaphones, and satellite navigation units. This is, of course, in addition to being a phone, and supplying messaging, email, and web browsing!

1.2.2 Smartphones and Feature Phones

A smartphone must be contrasted with a 'feature phone' which is a term commonly used to describe a low-end or mid-range mass market mobile phone which does not have such advanced functionality as a smartphone handset. A feature phone may offer a subset of the functionality of a smartphone, but is typically far more basic, with voice and messaging being the main features of the phone. Feature phone handsets are typically smaller, and since they have fewer features, need fewer electronic components; those used are frequently less technologically advanced. As a result, feature phones are cheaper to manufacture and to sell to the consumer; they are thus sold in significantly higher volumes (as the statistics in following section reveal).

---

[6] In 2006, Nokia was the world's largest digital camera manufacturer with approximately 140 million cameras sold through sales of Nokia smartphones and feature phones. It also sold close to 70 million music enabled devices, making Nokia the world's largest manufacturer of music devices as well. (Source: **www.nokia.com/A4136001?newsid=1096865**).

A feature phone does not allow for more than very basic personalization by the user through ringtones and themes or skins. Beyond these, only certain types of applications can be installed by the user, those that are sandboxed, typically written in Java ME. In consequence, the phones are said to be *closed*. Other restrictions may be in place, such as preventing access to web or WAP sites beyond those preconfigured by the vendor, or limiting the user's access to the phone's file system.

In contrast, the Symbian OS-based S60 and UIQ smartphone platforms are *open* because a user may install native applications written in C or C++, such as high-performance games, as well as applications written in Java ME.

So, besides more advanced technology found in a smartphone, one of the key differences between Symbian smartphones and feature phones is the ability to install native C++ applications. However, it's not quite as clear cut as this. One of the Symbian OS smartphone platforms, found in Japan, is also closed; it is not possible to install after-market software written in C++ on FOMA phones. For this reason, the FOMA smartphone platform will not be considered in the early chapters of this book where writing games in C++ using the native APIs and services are discussed. Those chapters are limited to S60 3rd Edition and UIQ 3. Since FOMA phones do allow games written in Java to be installed, Chapter 10 discusses writing games specifically for them using the DoJa standards, as well as discussing the Japanese game market. Chapter 9 is more general, and discusses how to write Java ME games for UIQ and S60 smartphones.

1.3   Some Statistics

1.3.1   Smartphones in Context

According to analysts at IDC, 528.3 million mobile phones shipped worldwide in the first half 2007. Of these, Canalys reports 47.9 million were smartphones. So it's clear that smartphones are only a small fraction (9 %) of all mobile phone handsets purchased. Having said that, sales of smartphones in the first half of 2007 were up by 39 %, compared to the same period in 2006. This can be contrasted with the more shallow rise of 17 % increased sales of all mobile phones over that time. (The continued momentum in mobile phone sales is driven by increased purchasing in emerging markets such as India and Africa, and purchasers in industrialized nations upgrading their current phones to the latest models).

Canalys forecasts that cumulative global shipments of smartphones will pass the one billion mark by 2012, while other analysts are less

H1 2007 smartphone shares by OS vendor, by region

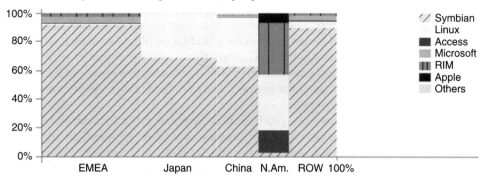

Symbian smartphone global shipments and market share

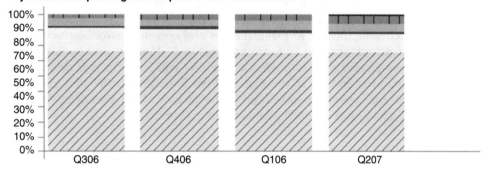

Figure 1.2 Smartphone market shares by region

cautious, and predict sales to break through one billion in 2011 (IDC) or in 2010. (Yankee Group).[7]

At the time of writing, as section 1.2 described, Symbian smartphones are reported to have a 72 % global market share. Regionally, the market share in China is reported to be 61 % (Q2, 2007 – Canalys figures) and, in Japan, the same research reports a Symbian smartphone market share of 72 %. Figure 1.2 reflects the market share, as reported by Canalys, at the time of writing (October 2007).

In February 2006, a Symbian press release announced a new pricing model for Symbian OS, designed to enable its licensees to target segments of the market for lower-cost devices, and further increase the sales volumes of Symbian OS phones. The scaleable pricing model reduces the price paid to Symbian per unit as the licensee's total volume of shipments increase. Using this model, Symbian OS royalties as low as $2.50 per unit are possible, reducing the cost of using Symbian OS in

---

[7] The Yankee group also predicts that smartphone sales will be 20 % of all mobile phone sales by 2010.

lower-price phones, and further accelerating the uptake of Symbian OS in the high-volume mass-market segment.

At the same time, Symbian also announced a collaboration with Freescale and Nokia to produce a 3G mobile phone reference design for S60 using a single core 3G chip. This reference design is expected to provide handset manufacturers with a cost-effective means of addressing the mid-tier 3G market segment and to reduce device development time by up to 50 percent.

Put together, the result is that Symbian OS can now be used to drive cheaper, mass-market phones as well as high-end smartphones. This effectively opens the possibility of deploying Symbian OS games to increasing numbers of consumers and leads to an interesting conclusion for mobile game developers: the addressable market for mobile games written for the Symbian platform is dramatically increasing in size.

"Yankee believes the primary growth driver for smartphones is the economic benefit to manufacturers and operators associated with standardized, scalable software architectures...Vendors and operators will enjoy rapid time to market at low cost with services and devices that span the entirety of the market, from basic phones to ultra-high multimedia centric models," said John Jackson, Vice President, Enabling Technologies Research, Yankee Group, 2007.

1.3.2   Sales of Smartphones vs. Sales of Portable Game Players

Let's take a look at some more statistics by comparing sales of Symbian smartphones with sales of comparable portable game players from Nintendo and Sony. As of 30th June 2007, Symbian reported 145 million cumulative smartphone shipments by its licensees. In the first six months of 2007 alone, 34.6 million units were shipped (a 44% increase on the same period for the previous year and more than for the whole of 2005).

Comparatively, Nintendo has published sales figures of 1.25 million DS portable game players in the first three months of 2007 (compared to 15.9 million Symbian smartphones over the same period), and over 47 million cumulatively since the DS launched. Sales of the Nintendo Wii since its release in 2006 were said to be just over 9 million units. Sony reports cumulative sales of its PlayStation Portable (PSP) to be 25.39 million units since its launch in 2005. Sales in the first half of 2007 were approximately 4 million units.

It's clear that there are a lot more smartphones on which to play games than there are portable game units. OK, it is also true that the PSP and DS are dedicated game players, and people that buy them are likely to buy one or more games to play on them, while not everyone who buys a mobile phone also buys mobile games to play on it. But, as Dr Mark Ollila, Director of Technology, Strategy and Game Publishing at Nokia puts it, *"With mobiles...the whole market is huge, so even if the percentage of those users playing games is quite small, the volumes are quite large.''*[8]

Recent data suggests that smartphone buyers show particular enthusiasm for purchasing and installing software for their devices. Part of the reason for buying a smartphone, after all, is that powerful after-market software can be installed onto it. To meet this demand, over 8000 third party installable applications are commercially available at the time of writing. Furthermore, according to data from m:metrics, Symbian smartphone users perform at least 20 % more tasks that raise the average revenue per user (ARPU) for network operators than other 3G users. For example, data from October 2007[9] suggests that, in the UK, 45 % of Symbian smartphone users spend more than £35 ($70 US) per month on 3G network services (such as sending photos, videos, accessing maps and news). Feature phone owners spend less; only 20 % spend that amount or more.

1.3.3 Sales of Mobile Games

The first, and probably best known, mobile game is **Snake**, found in most Nokia mobile phones since the Nokia 6100 phone was released in 1997. Since then, the mobile game industry has grown swiftly and is accepted as an important, and growing, sector of the game industry. We've looked at the statistics for sales of mobile phones in general, and smartphones in particular, so let's move on now to examine the market for mobile games in more detail. It's important to point out that this market data doesn't differentiate between games for smartphones and games for feature phone handsets.

Mobile games are a major revenue generator for network operators, game publishers and developers alike. For example, the market for mobile games took off in 2002, with the number of game services launched by network operators in Western Europe growing more than seven fold between mid-2002 and mid-2003, according to Screen Digest.

Various figures have been quoted for the value of the mobile game market to date, and for the years to come. iSuppli predicts that revenue is likely to increase rapidly, driven by heavy adoption of mobile handsets in China and India. Forecasts are for expansion to $6.1 billion by the year 2010.[10] A different report, from Informa Media & Telecoms, values the market at $3.3 billion for 2007 growing to $7.2 billion by 2011.[11]

However, it would be irresponsible not to sound a note of caution at this point. In 2003, the major market research firms predicted sales of mobile

---

[8] **www.mobileindustry.biz/article.php?article_id=2689**

[9] **www.mmetrics.com**

[10] **www.isuppli.com/news/default.asp?id=7242&m=1&y=2007**

[11] Informa Telecoms & Media: *2006 2007 Mobile Entertainment Industry Outlook*

games by 2006 to range from $3.6 billion (Informa) to $18.5 billion (Datamonitor); some were predicting sales of mobile games to be bigger than sales for dedicated game consoles and PCs combined. Although strong sales were reported in 2006, the market didn't take off as rapidly as expected, and Informa revised their estimate to $2.5 billion, which is generally taken to be accurate for mobile game sales that year. Later in this chapter, I'll discuss some of the reasons why the market wasn't as strong as predicted, and what is changing to address it.

1.4   Games Platforms Compared

Most people typically think of a game system as a dedicated game console such as Sony's PlayStation 3, the Nintendo Wii or the Microsoft Xbox 360. A console is most definitely not mobile! It does not have a display unit and is plugged into a separate television, and optionally a separate audio system.

The PC is another very popular game machine, particularly for games involving keyboard input, for example, to communicate with other players or characters in a game. Again, a PC is hardly mobile. If you play a PC game on a laptop, your play time is limited to the lifetime of a typical laptop battery, which is fairly short (shorter than the battery lifetime of a typical mobile phone). Furthermore, most PC games are not tailored for limited input from a laptop touchpad, but assume a mouse or joystick, as well as a full QWERTY keyboard.

Moving on to game machines that are more obviously portable, the most familiar at the time of writing are the Sony PlayStation Portable, the Nintendo DS and the Nintendo Game Boy series. While each of these have some level of connectivity (see Table 1.1), none have the facility to make voice calls from anywhere[12] or all the other characteristics of a mobile phone. This is one factor that differentiates a *portable* platform from a *mobile* platform. An additional difference between mobile and portable game platforms is that the mobile phone is ubiquitous; as we've seen from the sales figures, a wide cross-section of the population owns one and most carry them with them at all times. Relatively speaking, the market for portable game players is much smaller.

A portable game machine has fewer functions than a mobile phone, but does have the advantage that it is designed specifically to have controls and form factor that are optimized for game playing. Most mobile phones are phones primarily – game controls are a secondary consideration

---

[12] In early 2007, it was announced that PlayStation Portable owners in the UK would be able to make voice and video calls when using software developed by BT, through one of their BT WiFi hotspots. Nintendo DS users can use VoIP to chat during game sessions in some countries, but there is no commercial application to allow widespread use of the DS to make voice calls.

Table 1.1 Comparison of handheld game players and Symbian smartphones

	Sony PlayStation Portable	Nintendo DS	Nintendo Game Boy Advance	Symbian Smartphone (September 2007)
Graphics	High end 3D (approx 200,000 polygons per second)	High end 3D	2D	2D low end 3D (approx 25,000 polygons per second or lower)
Convergence	Camera and GPS attachment, Digital TV receiver attachment (available in some regions), Web browser	Web browser, MP3 player	MP3 player, e-reader bar code scanner	As described earlier in this chapter, Symbian smartphones are highly convergent
Connectivity	WiFi, IrDA, USB	WiFi, USB	WiFi, cable connection to Nintendo GameCube	WiFi, 2.5G/3G data and voice, SMS Bluetooth, USB, IrDA
Ergonomics	Dedicated game controller keys	Dedicated game controller keys, one touch screen	Dedicated game controller keys	No dedicated game controller keys (except in legacy devices such as the N-Gage and N-Gage QD game decks) • Standard phone keypad • 4-way directional controller (can be simulated 8-way in some smartphones) • Touch screen (some smartphones)

(*continued overleaf*)

Table 1.1 (*continued*)

	Sony PlayStation Portable	Nintendo DS	Nintendo Game Boy Advance	Symbian Smartphone (September 2007)
Support for independent games developers	Closed platform (there is some 'homebrew' development which exploits missing security functionality in early models)	Not officially supported, but it is possible to develop games independently	Not officially supported, but it is possible to develop games independently	Open platform, SDKs available from the UI platform creators (e.g., Nokia and UIQ)

(although the Nokia N-Gage game decks, discussed in Chapter 8, are a notable exception to this general rule). Another aspect for game designers to consider is that, because a mobile phone is a multiple-function device, the game may be interrupted at any time because one of the other services becomes active (e.g., the player receives a phone call or SMS, or a calendar alarm alerts them of their next meeting).

Not all mobile games use all the features that differentiate a mobile phone from a portable game player, but increasing numbers of games do use the connectivity offered by the phone network or WiFi. Some also take advantage of other features that are standard in mobile phones, particularly Symbian smartphones, but that are less well established in portable game players. These include features such as access to APIs to control the camera, microphone and location-based services. Chapter 6 discusses this further when describing some innovative mobile games created for Symbian OS smartphones.

Of course, while mobile phones offer a new set of opportunities for game design, there are also some trade-offs which lead to challenges when creating a mobile game. The mobile phone demands a very different style of game to those found on consoles. Games are played for a few minutes at a time; they need to be quick to grasp and deliver fun in short bursts.

A mobile device needs to be small and have a long battery lifetime, and is built with hardware limitations such as memory, screen size, and input controls to maximize portability and minimize price. Other, less obvious, limitations include how games are purchased and acquired. For mobile games, this tends to be different to the distribution of games for portable players, which are traditionally acquired as cartridges or on memory cards. We'll discuss some of the challenges of creating and distributing games for mobile phones in this chapter, and return to them throughout the book, since it's a general theme affecting mobile game design and implementation.

1.5   Types of Mobile Games

Mobile games have a broader, more diverse, audience than games for dedicated game consoles. Owners of consoles such as Sony's PlayStation 3 or the Microsoft Xbox 360 are keen gamers; they have purchased a machine specifically to play games, which are often expensive (when compared to typical mobile games), and they dedicate time regularly to playing games at home. In contrast, most mobile phone owners bought their phones for communication with others. The ability to play games is mostly a secondary benefit.

Most people who have a phone have at some point tried a game on it, if for no reason other than because it was already on the phone and they had time to kill.[13] A number will play games regularly – we'll look at the statistics shortly – but others will be more occasional users. Mobile games are cheaper than console games, and many will make impulse purchases of games, particularly if recommended by a friend.

In general, because the demographic of mobile phone owners is so diverse, the reasons for purchasing and playing games is variable, which means that the range of games available varies widely too. Mobile games are all somewhat limited by the form factor and resource limitations of the phones, but the range of games available is limited only by the imagination! Mobile games typically fall into two broad categories: casual games and hardcore games, with multiplatform games providing a bridge between mobile games and play on other platforms.

1.5.1   Casual Games

It's fair to say that simple games (often known as 'casual games') form a large part of the mainstream mobile game market. In fact, it is probably true for all games, not just on mobile platforms. It's been said that the biggest games in the world aren't **Halo** or **Zelda** or **Final Fantasy**, but the **Solitaire** game that comes with Microsoft Windows[14] and **Snake**, found on an estimated 350 million Nokia mobile phones.

Casual mobile games are targeted at a wide audience, and thus are designed to be accessible and easy to learn. Besides having simple rules, they only require the use of basic input controls (they are sometimes known as one-button games) and do not require long play times, so can be played briefly during a lunch break, on public transport, or covertly in a meeting. They are low maintenance. This kind of game is perfect for a mobile device. Most people always have their phones with them, so can use them to play when they have a few minutes to spare.

---

[13] Nokia's own research suggests that 68 % of Nokia phone owners have played the **Snake** game at least once, and given the number of **Snake**-enabled phones that the Finnish company have sold, this makes the game the most played in the world!

[14] *Edge Magazine*, May 2007

The nature of a phone is that it may receive an incoming call or message, to which the player will usually want to respond. Casual games are well-suited to interruptions, and are usually designed for the player to be able to return and pick up where they left off, or to be so casual that the player doesn't mind starting again (which is not true of more complex adventure or action games where a player works hard to move up to different levels, acquire skills, or gather resources).

Casual mobile games are usually free (for example, built into the mobile phone) or offered cheaply, because, by their very nature, casual gamers are unlikely to make frequent expensive game purchases. For this kind of business model to work, casual games are not as technically complex as hardcore games written for more dedicated gamers, which make heavy use of 3D graphics, audio, and have a more complex input, AI, and strategic gameplay. Casual games are also often limited in their installation size, because they need to be easy to download rapidly to the phone. This also limits the quality of the graphics and audio asset files that can be used.

The types of casual games available for Symbian smartphones include puzzle games (such as Sudoku and Tetris), card games (such as Texas Hold'em, Solitaire and Bridge), and board games (such as Backgammon, Chess, and Scrabble). Casual games are often based on simulations of games in the real world, because the rules are then already understood, and the gamer only has to pick up how the game is controlled. This is particularly useful on a mobile platform where the screen space available to explain the rules is limited. "*The most popular game we've created,*" says John Holloway of ZingMagic, "*is **Zingles** (Sudoku), a current worldwide favorite. People like it because it's easy to pick up and understand. It's also perfectly suited to a mobile device with only a numeric keypad.*"

1.5.2 Rich Content 'Hardcore' Games

The additional capabilities of Symbian OS, and the sophisticated smartphone hardware that it runs on, offer additional opportunities for game developers to take advantage of and write more complex games for so-called hardcore gamers. Smartphones allow for more sophisticated games using 3D graphics libraries such as OpenGL ES, or custom middleware and graphics solutions such as those offered by Ideaworks3D's Airplay Studio.

The Nokia N-Gage platform is designed specifically for developers of rich-content mobile games, those which follow in the footsteps of the games created for the original N-Gage devices (usually known as N-Gage game decks). The N-Gage game decks were intended for hardcore gamers and, during the lifetime of the handsets, Nokia published over 30 games with high-quality graphics, sound effects, and sophisticated gameplay. Chapter 8 discusses the history of the original Nokia N-Gage game decks and the new N-Gage platform, for professional game developers to create and distribute rich-content games.

On mobile phones, hardcore games need to take their lead from casual games, for example, to allow for shorter play times than the equivalent immersive console game, and to use clear graphics and simple controls. This is because the mobile phone is not a form factor for prolonged game playing and some phones cannot be expected to render highly complex images clearly, or have the ergonomics for input controls that require fast or complex combinations of key presses. This leads to the concept of a hybrid game, which takes the visual style of a console game but as a cut-down mobile version that can be taken on the move.

Ideaworks3D faced the challenge of creating a mobile version of **Final Fantasy VII** which combined the graphics and performance of the PlayStation 2 game but allowed it to be played on the mobile phone form factor using one hand. Says Thor Gunnarsson, Vice-President of Ideaworks3D: *"It's in the middle ground where a lot of the true innovation is. Probably we need a new way to describe it, maybe hard-casual or casual-core."*[15]

The Ideaworks3D solution is to make a tool chain similar to that used in console development but employing design techniques used in mobile. This enables artists and engineers from the console space to be able to produce content quickly and effectively. Besides that, and experience, plenty of play testing is needed to ensure the game is as engaging on a mobile device as it would have been on a console or PC.

1.5.3   Multiplatform Games

Leaving aside the creation of a mobile version of a popular console or PC title, such as **Final Fantasy VII** above, sometimes it may be more appropriate to make a mobile version which does not replicate the gameplay, but offers a different kind of interaction or view on the game. This is particularly apt for games that 'don't end,' such as massive multiplayer online role-playing games (MMORPGs). In these, gamers interact in a persistent virtual world hosted on remote servers so the game continues to run, and players drop in and out at will. Mobile versions of such a game may simply offer the player a chance to peek at their character's progress, or receive updates on some aspect of the game, perhaps by email or SMS, rather than play the game on the phone itself.

A good example is a game which, at the time of writing, is only known by its code name, **Project White Rock**. The game is currently under development by RedLynx, and is a multiplatform title expected for both

---

[15] BBC News: *news.bbc.co.uk/2/hi/technology/6445617.stm*

the PC and for N-Gage enabled S60 smartphones.[16] It is expected to be an innovative step forward for mobile games, and uses Nokia's SNAP technology, described further in Chapter 5.

Although not a classic MMORPG, the PC game *Spore* may offer a similar opportunity. At the time of writing, the game has not been released, but it has been described by the creator, Will Wright, as a *"massively single-player online game."* The information available suggests that the creatures and artifacts a player creates are uploaded automatically to a central database and then re-distributed to populate other players' games. The game would then report to the player how other players interacted with their creations. As of September 2007, it has been announced that *Spore* will be available in mobile format as well as for Windows PC, and potentially for other console and handheld systems. The mobile version of *Spore* is likely to be a good example of the future of multiplatform mobile games.

1.6 Who Plays Mobile Games?

We've already examined the statistics, which show that millions of people own a mobile phone, and that they are buying increasing numbers of mobile games. But where are they? What information is available about them?

It's actually harder to get statistics about mobile gamers than it is to get data about sales of phones and games. This is because mobile games are mostly downloaded from the network operators, who restrict access to this kind of data for competitive reasons. However, in February 2004, Sorrent and the U30 research company conducted a research study with 752 respondents, aged 9–35.[17] The data showed an even split between casual and hardcore gamers, and confirmed that there is more to mobile game playing than simply killing time on public transport (although this does continue to be cited by all mobile gamers as a regular use case).

The Sorrent research reported that over 60 % of the people questioned played mobile games at home for relatively long periods of time (15–20 minutes) and frequently (more than 65 % reported playing at least once a day).

A more recent report, with a larger sample size shows similar results. In 2006, Nokia commissioned Nielsen Entertainment to conduct research about mobile game playing in six countries worldwide.[18] Interviews were

---

[16] *blog.n-gage.com/archive/julyroundup/*

[17] *www.igda.org/online/IGDA_Mobile_Whitepaper_2005.pdf*

[18] *www.nokia.com/A4136001?newsid=1090119* summarizes the contents. The report can be found here: *sw.nokia.com/id/c52ab94e-e29d-498a-a36a-e80296e4184a/Evolution_Of_Mobile_Gaming_1_0_en.pdf* or simply by typing 'evolution of mobile gaming' into the site search tool at *www.forum.nokia.com*.

conducted with 1800 participants across China, Germany, India, Spain, Thailand, and the United States. The results found mobile phone gamers played mobile games for an average of 28 minutes per session – with some variations – India (39 minutes), United States (31 minutes) and Thailand (29 minutes). The respondents confirmed that they frequently play mobile games; the vast majority (80 %) playing at least once a week and 34 % playing every day.

The sample of people questioned by Nielsen entertainment was taken equally from those that describe themselves as either:

- mobile phone owners who play mobile games on their own, and with others (multiplayer games over Bluetooth or online)
- mobile phone owners who play solo mobile games, but rarely play with others
- mobile phone owners who do not play games on their phone, but do not reject the idea of doing so.

At the time of writing, the largest mobile games market is Japan, closely followed by South Korea, and then North America and Europe. The North American market is predicted to overtake that of Japan and South Korea in 2009[19] and it is predicted that the markets in China, South America, and India are where mobile games will take off rapidly (however, the price point for games in these regions will be significantly lower than in other territories). The reason for the uptake in India and China is that, in those regions, games consoles are relatively rare, while mobile phones are increasingly common. If you don't have a console on which to play, but want to play the latest game titles, it makes sense to go mobile.

1.7 Who's Who in Mobile Game Creation?

Having described who's playing mobile games, let's take a more detailed look at who is making them. This section will discuss the roles of game developers and publishers. In the following section, we'll then move on to discuss what happens when the game is ready to sell, and describe the roles of network operators, content aggregators, and independent channels in mobile game distribution.

1.7.1 Independent Game Developers

Hobbyist developers are people who write games for fun, perhaps to challenge themselves or learn more about a system. As Twm Davies,

---

[19] *Screen Digest* report by David MacQueen, May 2007.

creator of **Roids** describes in a paper on the Symbian Developer Network "*The game was originally written in one day (downing tools at 6am) as a challenge after witnessing a colleague take weeks and weeks to write an over-designed object-oriented asteroids clone.*"

You can find the paper Twm wrote about the experience at **developer.symbian.com/roidsgame**. Hobbyists tend to work on their own or in small teams, and there is a strong culture of open source and ideas sharing through developer discussion forums. The resulting games are often made available for free, although some hobbyists may sell their titles through independent distribution channels, or request a PayPal donation through their website.

From those writing mobile games for fun or a small profit, let's move on to those making a living writing mobile games. To cover all the aspects of professional game production, smaller independent game studios frequently need to outsource parts of the development process. This allows the core development team to cut down on expenses at times when they are not required. John Holloway from ZingMagic explains, "*We have a policy of outsourcing as much as possible on an 'as needed' basis. We use third party suppliers for many things like graphics, translations and play testing, and use services for accountancy and our website. This ensures we have as small a fixed cost as possible.*"

Independent game developers publish their games themselves, and distribute them through their own website, and through content aggregators, network operators and independent channels, each of which I will discuss in the next section. Because of the distribution model, independent developers need to consider how to help their customers when they have problems with the game. Customer support can be time-consuming, and take resources from future game development projects. It is important to anticipate the factors that may confuse an inexperienced user, such as security warning messages at installation time or complicated controls, and ensure they are either resolved or that help is provided in the game itself, or on the company's website.

Steve Townsend of Great Ape Software suggests providing a set of FAQs and a forum for players to discuss the game and any problems amongst themselves, with moderation from the development team to solve those issues that cannot be answered by the community. Great Ape Software is an example of a small independent developer team that uses its niche skills to create a compelling mobile game. Steve Townsend has over 15 years experience of kart racing, and has written a game using his knowledge that has received excellent reviews. More information is available at **www.greatape.com**.

1.7.2 Professional Game Development Studios

Most commonly, a professional mobile game developer works in a studio, which has a relationship with one or more game publishers. I'll discuss publishers shortly, so I won't describe that relationship here until each of the roles has been explained.

A professional studio may be separate from the publisher company or work in-house. For example, Electronic Arts (EA) is both a publisher and has numerous in-house studios worldwide, creating games for different platforms: console, PC, handheld and mobile. In comparison, Ideaworks3D Studio is an independent studio that collaborates with leading publishers to make flagship franchise games mobile, as well as to innovate original intellectual property (IP).

Game developers who work with publishers take responsibility for creating the initial game concept, technical design, implementing the code and creating any artwork and other assets the game uses, such as music and effects. The publisher's responsibilities are the subject of the next section.

1.7.3 Mobile Game Publishers

Game publishers fund the development of games, handle marketing and sales, and manage the development project overall. While the developer creates the game and does module testing and general testing as it is created, the publisher generally takes care of the game's play testing and QA. Game publishers, in both mobile and console/PC industries, usually acquire the rights to different intellectual properties (such as the license to create a game based on an upcoming film or associated with a particular sports star). They then set up a project to work with a developer team – either in-house or working in an external company. The game is usually ported for a range of different target game platforms. Examples of large mobile game publishers include Glu Mobile, Gameloft, I-play, Digital Chocolate and EA Mobile.

The publisher takes care of organizing the distribution of the game when it is complete, and builds relationships with the major network operators and content aggregators. Much in the same way that some publishers also have in-house developers, they may also be content aggregators. I'll discuss what services an aggregator provides later in this chapter.

The relationship between game developers and publishers is a close and symbiotic one. A failure to create a good game, or indeed any game in the agreed schedules, can have devastating results for both parts of the equation. The publisher will have approached network operators to promote the game, and will have promised it for a particular date, to a given level of quality. The network operators control distribution through their portals and only allow visibility to a certain number of games at a time, by allocating them slots on a 'deck,' which I'll discuss shortly. The games have to fit a particular genre, level of quality, and be well-received by the audience. If a delivery date is missed, or the operator receives complaints about the title, the operator will not pay the publisher. Additionally, the publisher may lose the coveted slot, and potential future slots, to a competitor, and may even lose credibility.

One Big Game – Game Publishing With a Difference

OneBigGame is a non-profit games publisher which aims to sell games to raise funds for children's charities around the world. Launching in Spring 2008, the types of games OneBigGame brings to market will be like any others out there, but they are created for publication by OneBigGame through the goodwill of game developers, in an industry-wide initiative.

Developers interested in doing something for the greater good with their creative skills can contribute skills and resources to help the publisher get games to market. The games are expected to be those that can be easily put together in a relatively short amount of time. The publisher is looking for donations, for example, any side-project that developers have been running to test a theory, or as a side-story to another game. The publisher takes the game, tests it and publishes it on the OneBigGame games portal, as well as through a variety of other online distribution partners. Any game that is fun and addictive, and that will sell well (and therefore generate money for charity) is welcome.

Mobile games are part of the OneBigGame portfolio, so I'm mentioning this initiative here in case any readers are interesting in getting involved. For more information, please see ***www.onebiggame.org***.

1.8   Mobile Game Distribution: Routes to Market

1.8.1   Network Operators

Network operators, or carriers as they are often known, are the largest vendors of mobile games; most mobile games are purchased and downloaded from their portals. The operators can be subdivided into international carriers, such as Vodafone, T-Mobile and Orange; national network operators – such as Cingular and Verizon in the US and DoCoMo in Japan; and other smaller carriers, such as mobile virtual network operators, whom I'll discuss shortly.

Network operators are very powerful in the mobile games industry because they stand between the game publishers and their customers. Customers typically do not select a network operator based on the games they offer, so the network operators can use their own policies to determine which games they provide and who supplies them. The games they select are displayed on what are called 'decks' on their mobile portal. The decks are simply catalogs of different categories of game, such as 'Puzzle' or 'Action' plus other promotional selections such as 'What's New' and 'Top Ten' bestsellers – a slot in these latter decks is highly desirable, because sales are typically better than if a game is categorized purely by genre. It's been reported that more than 70 %, and maybe even as much as 95 %, of the games purchased are from the main bestseller menus.

In general, the operators make commercial demands of the publishers in order to grant them a slot in the best decks. In the US, Alltel runs auctions for publishers to buy the top slots in the deck. In February 2007, it was reported that game publishers were prepared to pay up to 25,000 Euros ($34,000) a month to the French operator SFR. This payment secures regular top deck placements for two games a month, and partnering with the operator for co-promotion. The larger publishers have accepted this because they consider the up-front investment gives high returns, but smaller companies are unable to compete because they do not have the funds to pay for deck slots in advance of the game's sales.

For a game to be successful, the publisher or developer company must build a good relationship with at least one network operator in order to get the game selected for a slot in the profitable decks. Many of the large operators have so many publishers and developers wanting to work with them, that they do not engage with new contacts, but only use their established relationships with publishers and content aggregators to provide content.

The operators have individual processes for testing and certifying new games. They can each take a significant amount of time and, once a game is complete, a publisher can spend over six months submitting it for testing and negotiating the results with the operator, before the game is released on the operator's portal.

A mobile virtual network operator (MVNO) is a company that builds on top of the traditional network operators, because it does not have its own allocation of the radio frequency spectrum nor the infrastructure required to provide mobile telephone service. An example is Disney Mobile, owned by the Walt Disney Company, which offers voice service, handsets and entertainment content such as mobile games. The rise of niche MVNOs has been predicted, and since these operator types offer their own content distribution and billing, there may be scope for game developers that do not currently have relationships with established network operators to build partnerships with MVNOs that specialize in game downloads for a particular market segment.

1.8.2   Independent Channels

Some mobile game sales channels are not owned by network operators but are independent. These channels inevitably have less traffic than the operator channels, but game developers and publishers still find they offer potential for game distribution. Screen Digest estimates that about 5 % of mobile game sales are through independent channels.

There are a number of software download portals that provide mobile games as part of their offering, or specialize in them. Examples include Handango,[20] Motricity, Clickgamer, and MobiHand. These offer a web or WAP site for buyers to browse using their mobile phone and purchase games for direct download over the air (OTA), with payment made by means of sending a premium-rate SMS, which returns them a link from which to download the game.

Some purchases, such as those from Handango, can also be made over the Internet (OTI). The gamer visits the website on his PC, buys and downloads a game to it, and then transfers it to his phone, for example, using Bluetooth technology. Over the Internet purchasing offers the benefit of allowing the user to browse for games using the familiar medium of the PC, and find information about them, such as screenshots and reviews. The user can also download larger games more quickly and without needing the phone to be set up with a data connection, which is an advantage, particularly in regions where it is expensive, or slow, to download large amounts of data to a phone. OTI download also avoids being charged twice – once to purchase the game (by the channel), and then to download it (by the network operator). However, a number of network operators lock down their phones against what they call 'side loading,' which means that applications cannot be installed except by direct download to the phone. The lock down ensures that the operator retains some profit from application installation, even when it is not purchased from one of their channels. (Some go further and prevent the phones they distribute from making purchases *except* from their portals).

Even if it is allowed by the network operator, the downside of over the Internet purchasing is that the user must own, or have access to, a PC, and must know how to connect the phone to the PC to transfer the downloaded game and install it.

Purchasing from the web also requires electronic payment, for example using a credit card, rather than direct billing from the network operator. Not everyone has this option, especially when the target market for many

---

[20] In 2006, more than 250 new content providers and 3,500 new content titles were available on Handango for Symbian smartphones. Handango is a key channel for the distribution of after-market software for smartphones (based on Symbian OS, BlackBerry, Palm OS, and Windows Mobile) and regularly releases sales trends data, which it calls 'Yardstick Data,' on **corp.handango.com**.

games may be under the age of credit card ownership. The channels work around this through the use of gift certificates or vouchers.

Some phone manufacturers also provide sales portals, such as the Nokia Software Market. The portals are a way for manufacturers to make software available and encourage the uptake of their handsets. Rather than decide themselves what these sites provide, the portals often use the services of a content aggregator, such as Jamba or Handango, to put together the content offered.

The Nokia Software Market is available for users to buy games (and other applications) OTI (***www.softwaremarket.nokia.com***) or directly on the phone using an application called Nokia Catalogs. The Catalogs application is a client on the Nokia Content Discoverer system and is available for S60 smartphones and also Nokia's Series 40 feature phones. The application allows delivery of graphics, themes and ringtones, Java ME applications, native Symbian OS C++ applications for S60 devices, videos, music and content developed using Flash technology.

Sometimes, a handset may be configured differently depending on whether it will be sold via an operator or through a different channel. The operators configure the phone's look and feel to meet their branding requirements and may not include the Catalogs application. Nokia provides the Catalogs application for free download from their site (***catalogs.nokia.com***) for those who don't have it pre-installed but wish to use it to purchase content. When a phone is not sold through an operator, the application is pre-installed and ready to access the Nokia portal and partner channels.

Independent channels are usually the preferred route to market for small game developer companies because it is easier to place a game for sale with them than with a network operator or content aggregator. The portal will take a percentage of each sale of the game, but this is typically less than a network operator or aggregator, and the portal will allow developers to set the price of the game themselves. For example, the Nokia Software Market, takes 40 % of the sales price set by the developer, with revenue paid every quarter. For more information about working with the Nokia Software Market and Catalogs application, please see ***www.forum.nokia.com/main/software_market/index.html*** or consult the Nokia Sales Channels section on the main Forum Nokia developer site at ***www.forum.nokia.com***.

1.8.3   Content Aggregators

Content aggregators are specialists at gathering content, such as mobile games, applications and ring tones. The content is acquired from suppliers, such as publishers or independent portals, and distributed by the aggregators to their sales channels. Content aggregators provide an easy way for businesses to source downloadable material without having to seek it out and build a relationship with each supplier. An example of a content aggregator is Jamba (also known as Jamster in English speaking regions), which is one of the largest channels for mobile content and applications. Jamba also distributes content and applications via the Nokia Catalogs client on Nokia devices, which was described in the previous section. Forum Nokia publishes a comprehensive list of mobile content aggregators at **www.forum.nokia.com/main/go_to_market/aggregators. html**.

Some content aggregators can be publishers too, and this can cause problems to independent developers who do not work with that particular publisher. If there are two similar games: one from an independent source, and the other published by the content aggregator, it's clear that the aggregator is most likely to opt to promote his or her own game. Some aggregators are also channels in their own right – such as Handango.

Confused? It's probably obvious by now that the mobile games ecosystem is full of complex, and occasionally incestuous, business relationships. Breaking into it requires a good understanding of the different roles, and keeping track of the changing ownership of the publishers, channels and developer studios. An interesting history and summary of how the present day mobile game market evolved can be found at **www.roberttercek.com** (look for the links to the GDC 2007 presentation). As a summary, Figure 1.3 attempts to show some of the roles and routes to market.

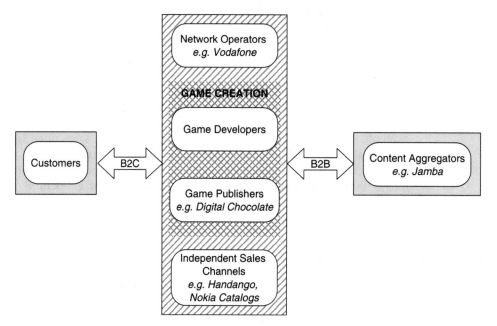

Figure 1.3 Some of the business-to-business (B2B) and business-to-consumer (B2C) relationships in the mobile games ecosystem

1.8.4 Built-In Games

One way to distribute your game to users while avoiding the complicated relationship between publishers, content aggregators, and network operators is the holy grail of game developers – getting the game built onto a phone handset by the manufacturer, (or provided on a CD ROM in the box that contains the handset). Naturally, there are far fewer opportunities for built-in games than there are developers who wish to do a deal to get their game on a phone. From a developer's perspective, the market exposure of an embedded game is an excellent opportunity; however, it is not without its problems. Trying to ship a game on a handset that has not been released means the developer is often working with pre-commercial firmware and hardware. Final versions of the hardware may become available just days before the deadline for embedded software on the handset, resulting in uncomfortable 'crunch' periods for developers.

When a game is built into the phone, it is usually available to play for free for the lifetime of the handset. Some games distributed on the in-box CD may have a limited duration, after which they must be purchased by registration. The embedded demo contains instructions on how to purchase a full game and continue play. One example of this is Ideaworks3D's **V-Rally**; a demo version of which was shipped with the Sony Ericsson P900i.

A developer is usually paid up-front for delivering a game to be built into the phone, rather than paid a royalty per handset sold with the game on it. Additional payments may be forthcoming if the game is put on further handsets, but the game, including the source code, usually becomes the property of the phone manufacturer.

Sometimes a network operator will customize a phone after it leaves the manufacturer, and put certain games, ringtones, themes and other pre-installed content, onto the handsets before distributing them. Developers chosen to provide a game to the network operator for this purpose will retain rights over the game itself, although the terms of the deal will vary, they will also typically receive a one-off payment rather than royalties based on handset sales.

1.9 The User Experience

1.9.1 Finding a Game

Many users report that the mobile download medium is not a particularly good one for browsing games or finding out more about them. For a start, they can sometimes be hard to find at all on the basic service menus. Network operators have other attractive content to distribute besides games, such as music, video or mobile TV. Therefore, the promotion of mobile games can be limited, and a user may have to actively search the site – which can be frustrating when they are using a phone with a slow data connection and a small screen.

On the topic of limited screen real estate, even when the games are easily located, the phones allow only a few lines of text to be displayed. Console games sold in retail outlets have large boxes with space for artwork and text to give a potential purchaser an idea of what the game is like. Mobile games distributed for download over the air do not have much space to do this. Besides the attractive pricing, this may be part of the reason why casual games are so popular – the buyer already knows about **PacMan** or **Tetris**, so he or she can be confident about understanding what the game is about when it is downloaded (though it doesn't give any prior knowledge of how to play it, or guarantee of how good it will be). A user perceives that it will be difficult to get customer support for the game, and since it doesn't come with a box of instructions, he or she will generally have to work out how to use it by trial and error, or using the help system built into the game, which again is limited by the screen size available.

On the operator portal, a potential purchaser may also find it difficult to find or share reviews, or hints and tips about a mobile game, which are key elements for many gamers. The combination of a relatively slow browsing speed and small form factor make browsing for information difficult, unless the comments are made directly available on the deck with the game. The role of the community in finding and recommending games, and sharing experiences and reviews is something that the Nokia N-Gage platform emphasizes.

The constraints of the size and information on the decks are also thought to be the reason why many consumers just buy games from the bestseller decks on the operator portal. The previous sales give more confidence in the game, and the bestseller games are easy to find on the deck.

1.9.2   Billing

The poor user experience associated with purchasing a mobile game is widely cited to be one of the main reasons that sales did not take off as rapidly as predicted. Besides the frustrations of trying to find the game they want, users can also be confused about billing, particularly if they are charged once for the game itself, and then extra in the tariffs applied when the game is downloaded.

To generate additional revenue beyond a one-off payment when the game is purchased (or as an alternative to it), mobile operators have introduced a number of different mechanisms for billing the consumer, including subscription billing and premium SMS billing. Games that generate great customer loyalty can attract a monthly subscription just to allow the gamer to continue to play them. Others are episodic, whereby additional storylines are released on payment of a regular subscription. This is similar to the concept of an expansion pack which may be purchased by subscription, or just occasionally. Expansion packs provide add-ons to the original game, such as new tracks for a racing game, team members for a soccer game, or weapons for an action game.

The subscription model is most common in Japan, where it is part of a generic portal-based mobile Internet service called i-mode. Mobile gamers pay a monthly charge and can subscribe to a single game or can download and play a number of smaller games each month. Chapter 10 explains more about the success of i-mode and the market for mobile games in Japan. In Europe, i-mode was introduced in 2005, but uptake has been poor; for example, the UK operator O_2 reported in mid-2007 that it would phase out its support for i-mode by 2009.

According to one game developer and publisher in Europe, *"Sub-scriptions terrify the consumers"*.[21] The payment model has not been popular for various reasons, notably the lack of information about how to unsubscribe, billing practices where users may not even realize they are signing up for a subscription, and dubious content quality.

An alternative to a subscription is to allow players to decide when they want to purchase additional units to extend or continue playing the game. In the US, this is increasingly popular, and is sometimes known as 'rental.' The operator can set a low price for a one-time purchase of a game, but can then additionally combine it with the rental billing model to generate additional on-going revenue.

1.10   Mobile Game Platforms

When writing a game for a mobile phone, there are various technology platforms upon which the game is built, chosen according to the type of game, the performance it requires, and the phone handsets it is intended to be played upon. A technology platform comprises a set of the software components to access the mobile phone hardware application environment, and varies according to whether it interprets the game code or compiles it to native instructions – and if so, it additionally depends on the operating system used by the hardware. Let's look at the most common mobile games platforms –Java ME (which is supported by all Symbian smartphones) and BREW (which is not) and contrast these with native game development for Symbian OS in C or C++.

1.10.1   Java ME

Java applications execute on a virtual machine (known as a 'VM') that runs on top of the native platform. Applications are effectively hardware

---

[21] *Issues in the Mobile Game Market*, DFC Intelligence, March 2007.

agnostic, since they are released as Java bytecode, which is portable between different processor architectures. The bytecode is interpreted by the VM implemented for that specific platform, and applications are translated into native hardware processor instructions. The translation does lead to slower execution rates for Java code than the equivalent code written in the native instruction set. However, there are acceleration strategies adopted, both in any specific implementation of the Java VM, and in the bytecode that the VM interprets and executes. Java is also a robust technology – Java instructions cannot cause the system to crash and security features are designed into the Java programming language and VM.

Games for mobile phones that are written using Java actually use Java ME (the 'ME' stands for Micro Edition). This is the version of Java for mobile devices, that was previously known as J2ME (Java 2 Platform, Micro Edition) and it offers a subset of the full Java platform for resource-constrained devices such as mobile phones. Java ME devices implement a profile – for mobile phones this is called the Mobile Information Device Profile (MIDP). Applications written for this profile are called MIDlets.

Java ME was not originally designed as a games platform, and it initially lacked some of the functionality needed by games, such as graphics effects like layering, transparencies and sprite rotation. Some handset manufacturers added their own interfaces to supply this and other 'improvements' and, combined with some loose, or optional, specifications, the Java ME platform became somewhat fragmented. In effect, games were not portable between phones from different manufacturers and required effort from developers who found themselves producing multiple different versions for different phones (sometimes over 500 versions for a popular title!). This was a long way from the 'write once, run anywhere' model for Java, and caused many problems, although the handset manufacturers quickly realized that it was hindering development for their devices, and they are now moving to standardize their implementations.

Working groups, called Java Specification Request (JSR) Expert Groups, design new APIs for mobile devices – to allow the Java ME platform to evolve and adopt new technologies. As the number of JSRs increases, the platform is becoming increasingly capable of providing support for high-quality games.

Each proposed API is described, discussed and eventually approved, at which point, it can be implemented by any mobile handset manufacturer wishing to do so. Each implementation of an API must be carefully tested against the JSR for compliancy, and a game that exploits a particular API should then be portable across devices which offer the same JSR. In spite of conformance tests, implementations of some features may still differ according to the handset or manufacturer in question. Subtle differences can creep in, because of the implementation differences on different hardware, in some cases, even across device families from the same manufacturer. However, despite some remaining fragmentation, and with some limitations, Java ME code is mostly portable across various operating systems and processor types.

Most mobile phones – Screen Digest estimates the figure to be over 90 % – support the installation of Java ME games. For a number of lower-end phones, and for Symbian smartphones in Japan, it is indeed the *only* route to use to deliver games to the phone after it has left the factory. Because of its ubiquity, Java ME is the largest technology platform for mobile game development. One of the advantages of developing in Java ME is that Java is a popular programming language and has good tool chain support and documentation. Developers have access to multiple well-established programming resources, and companies have a large body of skilled developers to recruit from.

Chapter 9 discusses the creation of Java ME games on Symbian smartphones in detail, and Chapter 10 describes the DoJa standards for developing games in Java for the Japanese market.

1.10.2 BREW

BREW (which is an acronym for Binary Runtime Environment for Wireless) is a proprietary runtime platform created by Qualcomm Inc. BREW can be supported by handsets built for GSM/GPRS, UMTS or CDMA. However, when BREW was first introduced it was solely developed for CDMA handsets.

While Java ME games are delivered as bytecode that is interpreted by the VM rather than associated with a particular processor, BREW applications are compiled into ARM machine code. The compiled code is said to be native and, as such, BREW offers an improvement in performance. The BREW runs between the application and the phone's operating system, so applications do not need to use system APIs directly. BREW developers write their code in C or C++ using the SDK provided by Qualcomm.

The BREW SDK is free, but developers must pay for support if they need it. Unlike Java ME games, BREW applications have greater direct access to the hardware so they must be digitally signed before they can be deployed to a phone, to ensure that they do not compromise its behavior, either maliciously or accidentally. Once an application has been created and tested internally by a game developer, it must be submitted for independent testing (known as 'TRUE BREW') to ensure a defined level of quality. Only then can it be distributed commercially. This process can increase a game's time-to-market and the testing and certification process is an additional development expense. However, it is also true that many network operators insist that a Java ME game be tested and certified before they will allow it to be distributed via their portals. So, although the requirement for TRUE BREW testing may put hobbyists off working on the platform, it does not usually deter professional game developers, who find that rigorous certification, while time-consuming, is a *de facto* part of any game development process.

Qualcomm collects royalties from the developers who deploy code on the BREW platform. In return, it provides a set of consistent and rich APIs for BREW that includes those for distribution of the application binary to the end user. The provisioning system integrates with network operators' billing systems to make it easy for users to make purchases, and easy for network operators to track purchases. The model also allows variation from one-off payments for a game, and provides, for example, a subscription model for regular revenue generation.

BREW is considered to be an excellent platform for game development, particularly BREW 2.0 which has good support for 3D graphics, and can be used with powerful handsets to create high-quality games. The end-to-end system is considered consistent and attractive to many games developers, and because of the power of compiling to native code, it is a very popular choice of platform.

BREW is mostly used in North America, China, Taiwan and South Korea; globally, *Screen Digest* estimates the install base of BREW phones to be approximately 6 % of the total market. Even in the markets where BREW is available, evidence suggests that the majority of mobile games purchased are Java ME.

We won't discuss BREW any further in this book, because it is not a platform offered by Symbian smartphones.

1.10.3   Native Platforms

The term native in this context refers to games written in C or C++ for a particular mobile operating system, such as Symbian OS, Windows Mobile or Palm OS. The games are programmed against software APIs that are provided in libraries found in software development kits released by the phone manufacturer or platform vendor, many of which originate directly from Symbian. Alternatively, they can be written to use a layer between the game and the OS, such as 3D graphics middleware, the N-Gage platform or the standard C libraries, provided by Symbian, Nokia and other partner companies.

The benefits of writing native games are that more direct access is granted to the hardware of the phone, which improves runtime performance and flexibility to manipulate the phone hardware. For example, a native graphics library is significantly more capable and efficient than its equivalent in Java ME.

In consequence, native games are typically more sophisticated than Java ME games, with better visual and audio quality, and often more innovative use of the phone, because access to certain features, such as motion detection, vibra and camera may otherwise be limited by the JSRs a phone supports, and the scope of the JSRs themselves.

Some game developers prefer to work in C or C++ rather than in Java ME, even when creating basic, so-called casual, games that do not need access to the 'bare metal' of the hardware or high performance graphics. One reason is that they can license a particular library to use in their game that was created in C or C++ or they may have their own legacy game engines and wish to re-use that code.

However, commercial developers often find themselves competing with hobby developers who are writing similar games in Java ME and giving them away free or selling them very cheaply. Unless you are an experienced C++ developer with an established game engine, architecture and test regime in place, the effort involved to create and deploy a native casual game is usually higher than to create a rough equivalent of the game in Java ME. Many professional developers will not find the additional effort worthwhile when competing for sales of cheaper games.

Portability is an issue for games that compile to native code, because the games will only run on particular handsets. The code must be compiled for the instruction set according to the processor in the phone and against the set of software APIs that a particular operating system provides to interact with the hardware. For example, a C++ game written for a Windows Mobile smartphone will not run on an S60 smartphone built upon Symbian OS, much as an application for a Mac will not run on a PC. The game's source code must be re-compiled and, before this is possible, it usually requires significant code changes because of the differences between the operating systems.

Experienced game developers design their games to separate source code that is generic and can be re-used on each platform from the code that is specific to a particular operating system. This makes porting a game which compiles natively as easy as possible, although it still requires code development, integration, recompilation and testing for each additional platform.

Says John Holloway of ZingMagic, *"We have as much shared code as possible, even at the UI level. Drawing the correct chess pieces on a chess board in pseudo Z order to achieve a 3D type effect has nothing to do with Palm OS, Windows Mobile, Symbian OS or any other GUI specifics. We can build a generic logical model of what we need to do then map that into real platform function calls for each. That way we can write much of our code for multiple platforms simultaneously."*

There are also a number of standard APIs that abstract API differences and allow mobile game developers to re-use the source code for a game without making any, or significant amounts, of changes. These include standards for graphics (such as OpenGL ES which is an API for fully-functional 2D and 3D graphics) and for audio (such as OpenSL ES). The C and C++ standard libraries are also a means to abstract native platform differences, which are may otherwise require source code changes before recompilation. These can be particularly useful on Symbian OS, which has a set of idioms that are unfamiliar to many experienced developers who have previously worked on other platforms. Chapter 7 discusses the support for various standards that is offered by Symbian OS.

Aside from the portability between different native mobile platforms, another drawback is that the installation of native games is often restricted or prevented. Many network operators do not allow the download and installation of native games, and prevent the user's access to the file system of the phone except through specific application managers that do not recognize the file types of native installation packages. Many lower-end, mass-market phones, and Symbian smartphones in Japan, are closed to after-market installation of applications except those written in Java ME or DoJa, according to region.

To conclude this section on technology platforms, let's set out where you can find more information in this book. We discuss the C++ APIs supplied by Symbian OS for creating native games in Part Two of this book, standards support for games creation and porting using C/C++ in Part Three, and how to write games for Symbian smartphones using Java ME in Part Four. On S60 3rd Edition smartphones, mobile games can also be written for Adobe Flash Lite, which is discussed further in Chapter 11.

1.11 Portability and Compatibility

Portability and compatibility of mobile games are big issues inherent to the lifetime of a typical mobile games device, and the number of different devices available. Let's compare the expectations of a console game developer, a PC game developer, and a mobile game developer.

1.11.1 Consoles

The developer of a console game knows that if he or she tests the code on a Microsoft Xbox 360 before releasing it, it will run on a customer's Xbox 360 too. It will also run on the customer's friends' older Xbox 360 consoles, and, in two years time, it will also work on a newer Xbox 360. The only reason that it may stop working is if compatibility is broken by the hardware manufacturer, for example, if a firmware update is introduced that has not been tested for backwards compatibility. This is likely to be rare because the manufacturer does not want to fragment the platform into different firmware versions since this causes anxiety for consumers who can't easily tell if a game will work on their particular machine. Indeed, some of the console manufacturers go to great lengths to ensure that, not only is a single console compatible throughout its lifetime, but that it also plays games from previous consoles in the series. For example, a software emulator can be used on an Xbox 360 to allow original Xbox games to be played on it. Likewise, the Nintendo Wii is backward compatible so that all official Nintendo GameCube software may be played (although a GameCube controller is required to play GameCube titles, and online features of those games are unavailable on the Wii).

There is a very limited concept of portability in the console world. There is only one model of a console for a developer to worry about – subject to minor variations such as hard disk space or memory.

Of course, *between* consoles there is a portability issue. A console game written for Xbox 360 cannot be run on a Nintendo Wii, because the two platforms are *not* portable, but most users understand that. A game must be ported to run on different consoles, and large game publishers often do produce ports to allow popular games to be played on multiple different console platforms. A good game developer, regardless of whether they are working on console, PC or mobile platforms, will usually design a game to make the porting effort between devices as easy as possible.[22] We'll also discuss some of the standard and middleware used by game developers to make their code more portable in Chapter 7.

1.11.2 PCs

A PC game developer has more issues to consider than a console developer, since the game must be tested on each of the different versions of Windows available, to ensure the game's portability between them. For example, a player may have Windows XP, but then upgrade his or her PC to Windows Vista. He or she will expect to be able to install and run the game on the different OS platform if the manufacturer claims it is supported. There may also be issues associated with hardware differences between different PC setups, particularly since PCs can vary significantly in their capabilities. The PC game developer has to consider these when creating the game, and list a minimum set of capabilities as part of the game's specification.

Compatibility is also not a problem, since Microsoft provides a level of assurance that newer, successive versions of Windows will retain compatibility with older releases, for a timeframe which is typically longer than the lifetime of the game.

1.11.3 Mobile

Mobile game developers face the biggest challenges in terms of portability and compatibility. There are numerous different phone models in the

---

[22] For more information about the issues to consider when writing portable games, I recommend *Cross-Platform Game Programming*, Steven Goodwin, Charles River Media, 2005.

market, from different manufacturers, each with varying hardware and different operating systems. This is typically called *fragmentation*.

I discussed issues of portability for native mobile games in the previous sections. For example, Java ME developers have to be aware of device fragmentation. Different handsets may support different versions of the Java ME standard and different sets of JSRs. To work around this, the game developer has to create separate versions of the game for the different phone types, sometimes numbering into the hundreds for a popular game, and increasing the development expenses caused by testing across the range of handsets supported.

Even when a developer manages to resolve the technical issues of portability between different phone handsets so that the game executes, it doesn't necessarily mean that it will be easy to play! Handsets may vary significantly in terms of screen size and orientation, memory specifications and input controls. The difference in the form factors between mobile phones can be significant, compared to that between PCs (where each is likely to have a mouse, keyboard, and large screen). For example, some phones have keypads with reasonably-sized, well-separated keys, a joystick and high contrast screen. Others may have small, stylized keypads, touchpads or small screens. Some of the current Symbian smartphones are shown in Figure 1.4.

Figure 1.4   The diverse range of Symbian smartphones

A game designer often has to decide on a basic common set of hardware features, and support those alone, in much the same way that a PC game developer presents a minimum set of hardware specifications.

Unlike consoles, which generally have a lifetime of about five years, the typical mobile phone lifetime is a year to eighteen months. Mobile technology evolves rapidly, driven by the dynamics of the mobile communications business and consumer expectations. Network operators actually subsidize the purchase of each new generation of phone handset in order to spur consumer adoption and drive up the use of new network services and tariffs.

1.11.4 The Symbian OS v9 Binary Break

The underlying mobile platform changes according to the requirements placed on it by operators and consumers, and in doing so compatibility may not be retained. A good example of this is the binary break introduced by Symbian OS v9, which is found in UIQ 3 and S60 3rd Edition. Applications for earlier versions of those UI platforms cannot be installed and run on newer Symbian OS v9 phones because of binary incompatibility across the Symbian platform. The code must be recompiled, and requires changes in all but very few cases.

Symbian and its phone manufacturer licensees do not break compatibility lightly, and do not plan to do so often, but could not avoid doing so in Symbian OS v9. The migration path between the original and new platform versions has been clearly documented for developers, but it is still clearly an issue that needs to be highlighted. This book will focus on writing games for Symbian OS v9 only, and will not describe the earlier platform versions. For a detailed discussion of the differences in writing code pre- and post-Symbian OS v9, please consult the porting documentation on each of the main developer support websites listed in the References and Resources section at the end of the book.

1.11.5 Symbian Smartphone UI Platforms

There is a further issue for developers targeting Symbian smartphones that should be discussed at this point. As described earlier in the chapter, depending on the handset manufacturer licensing Symbian OS, the smartphone uses one of the three UI platforms: either Nokia's S60, NTT DoCoMo's MOAP user interface for the FOMA 3G network, or UIQ, designed by UIQ Technology.

There are significant differences among the three platforms. At a high level, these can be characterized in terms of look and feel of the UI, and the way the user interacts with the phone. For a developer, the distinction occurs at an API level – code for one UI platform is incompatible with the other platform. Developers using native C++ APIs find there are some significant differences between UIQ 3 and S60 3rd Edition, in the UI classes employed, how they are used, and the design philosophy employed. Code conversion between them can be complex. It can lead to almost two completely different versions of the application, unless the application is carefully designed to separate generic code that uses standard Symbian OS APIs, such as string, event or file handling, from code that displays the UI and code that accesses hardware such as the phone's backlight or the vibra motor that vibrates the phone.

Fortunately, games are often some of the easiest C++ applications to convert between UIQ and S60. Many games do not use the standard UI controls (such as dialogs and list boxes) that other applications employ, and thus typically require fewer code changes. Of course, games often use hardware specific APIs and these need to be ported between platforms. Some may use elements of the UI platform in their menu system, although many define their own. Good game design will separate platform specific code from agnostic code, making cross-platform conversion relatively straightforward.

Even within a single UI platform, there can be issues of portability. The hardware characteristics of a range of smartphones can differ significantly according to the target market for a phone. For example, Nokia's S60 range offers fully featured 'multimedia computers' which are smartphones featuring, for example, high resolution screens, hardware accelerated graphics, high-quality video and WiFi, for downloading mobile TV, music and video. Then there are alternative, smaller form factors, and more basic S60 smartphones for those that want the benefits of a smartphone (such as email, web, and ability to install after-market software) without wanting all the multimedia options available. Variations such as the amount of memory available, processor speed, and disk space are also factors that may affect some games, for example, if the frame rate is crucial or there are a lot of game assets to store and load into memory.

Many games aren't significantly affected by variations in the phone's processor, memory, or storage, but variations in the form factors may cause issues. Some phones may not have a camera, for example, or may have a lower resolution screen than others. A game must be one of the following:

- able to determine the device characteristics and adjust accordingly

- optimized for the lowest common denominator hardware it can be installed upon

- agnostic to any variation in hardware.

The N-Gage platform addresses the issue of portability between different S60 smartphones by providing a set of libraries that present a consistent interface for the games to establish hardware capabilities and manage differences. The N-Gage platform approach also divides the S60 phones that support the N-Gage platform into 'device classes' according to capability, so a game developer can guarantee a minimum level of functionality, for example, a minimum allocatable heap size of 10 MB and a minimum storage capacity (on external memory card or built-in memory) of 64 MB, and support for standards such as Bluetooth 1.2 and OMA DRM 2.0.

The N-Gage platform also allows game developers to write games for Symbian using C++ that is more familiar, using standard interfaces and types, rather than relying on knowledge of active objects, descriptors and Symbian OS-specific types. This in itself eases portability of game code originally written for other platforms, such as Windows Mobile, and even console or portable game players. Chapter 8 discusses the N-Gage platform in more detail.

Some game developers use middleware solutions to give them portability across platforms and devices. A good example is Ideaworks3D's Airplay. The Airplay SDK provides an operating system abstraction layer, 'Airplay System,' which abstracts all of the major mobile operating systems. This layer not only provides source level compatibility but also goes a step further by maintaining binary level compatibility between platforms, which results in major cost benefits (for example, QA time is significantly reduced, as every platform is running the same game application binary). In addition, the publisher is able to deploy to all platforms simultaneously, which is critical to being able to leverage their marketing expense for a given title (often tied in to a console game or film release). Some of the middleware solutions available for Symbian smartphones are described in Chapter 7.

1.12   Smartphone Characteristics

The characteristics of a mobile phone differ from a typical handheld, PC or game console, and this has an effect on the design and implementation of mobile games. As I've already discussed, a Symbian smartphone is a bit like a Swiss army knife, and has a number of different roles. A game developer must take this into account; the code must co-exist with the other applications running on the smartphone and take cues from the operating system when necessary.

For example, a game should be interrupted when there is an incoming call, and should allow the player to be aware of received messages and display calendar alarms, if he or she chooses to be notified while immersed in the game. The operating system controls much of this and will, for example, force the game to the background when a phone call is received. The game is made aware of the change in focus, and the game developer should write code so the game suspends its activity while it is in the background – we'll discuss how to do this in the next chapter. Game developers must take measures to prevent the game consuming battery power and other resources, such as active network connections, while the game is paused, particularly if it is for a lengthy duration.

More generally, it is important that a mobile game developer considers the limited resources available on a typical Symbian smartphone compared to a PC or game console. Good examples are the limitation on memory, disk storage capacity, and download size, and the fact that it mostly runs on battery rather than on main power. This last factor, combined with the standard ergonomics of a phone and the context in which mobile games are played, typically limits the time for which a game is played. Mobile games have much shorter play times, in general, than their console or PC equivalents, and a game designer must consider how to grab a player's attention, and be satisfying, in a short time frame. The need for simple, consistent gameplay and the circumstances under which mobile games are often played (in transit, with associated issues of visibility, sound constraints and interruptions) are important issues in the design of any mobile game.

A game designer must also take into account the variety of form factors such as input controls, screen sizes and screen orientations available to the mobile phone platform. One interesting issue is that of localization, where a game is translated into a number of different languages. The game may be designed with artwork and text for one particular language, often English, and the layout arranged to fit the small phone screen exactly. Text localization may result in different length strings which are often significantly longer than the English version, but still need to be made to fit in the same screen area. Artwork may also occasionally need to change for locale variations, perhaps for language or cultural reasons, and the limitations of the screen real estate must be considered in advance.

Similarly, many games are designed for play on landscape orientation screens, particularly those for mobile game players such as the Nintendo DS or Sony PSP. Most mobile phones have portrait aspect screens, although some Symbian smartphones do have landscape orientation screens or the ability to switch dynamically between either layout. This is an important factor when designing the graphics of a game, or determining how to port the game from another platform.

1.13   The Future for Games on Symbian Smartphones

What is changing to improve the user experience and drive the uptake of mobile games on Symbian OS to greater sales volumes?

1.13.1 Easy and Cheap Downloads

Most people keep their smartphones with them at all times. If it's easy to find good games, they are likely to download and play them. Ubiquity is an advantage of the mobile platform – but until now, the expense and difficulty of finding games has put a brake on purchases.

The rollout of next generation high speed data networks will be an important factor in facilitating download of larger, better quality mobile games. This combined with fixed-rate data plans – and the increasing prevalence of WiFi in Symbian smartphones, such as the Nokia N95 and E61i – means that purchasers can make more regular downloads and game purchases. It can also be expected to boost quality, and particularly the market for the rich content native games supported by Symbian smartphones, such as those for the N-Gage platform. Games with good visual and sound quality need sophisticated and large game asset files and would otherwise be too expensive, or impossible, to download over the air.

For 2.5G handsets, the maximum download should be limited to 500 KB for a good user experience. For 3G networks it is acceptable to support up to 2 MB downloads. For 3G High Speed Access, UWB, CDMA EV-DO, or WiFi networks, larger downloads (e.g., 10–25 MB) should be acceptable.

A nice example, given in 2005, is the *"cycle of improving quality, driving and being driven by improving technology ... seen in the evolution of the Internet. As more people gained broadband access more complex content became readily available. As more content was offered, more people wanted broadband access."*[23]

The importance of making it easy for a user to discover, try, play, buy, and share games is something that the Nokia N-Gage platform emphasizes with its N-Gage application for S60 smartphones. The N-Gage platform launches at the end of 2007, after this book goes to press, and the future uptake of mobile games on Symbian OS smartphones may well be heavily influenced by the user experience it introduces.

1.13.2 Game Quality Improvements

A guaranteed level of quality in the experience of purchasing a mobile game is needed to improve the uptake of mobile game purchasing. At the 2007 GC Developers conference in Leipzig, Diarmuid Feeny, Nokia's Games Business Manager, quoted statistics from Nokia's channels that show 50 % of mobile phone owners are willing to buy games, but less than 5 % actually do so, and then less than half of them make a repeat

---

[23] *www.igda.org/online/IGDA_Mobile_Whitepaper_2005.pdf*

purchase. It seems that the potential customer base is wary, and the reasons cited for this include the experience of making the purchase and the unpredictable quality of the game.

The technical constraints against writing good quality games for Symbian OS are few. The capabilities of Symbian smartphones are constantly growing, offering larger screens, faster processors, and more memory in each new generation of smartphone. Some smartphones, such as the Nokia N93 and N95, have hardware acceleration, allowing console-quality graphics. The increase in memory and disk space available on Symbian smartphones supports the enhancements in download speed described previously, allowing larger assets for higher quality gameplay.

The increasing support for standards and game development middleware that Symbian OS and the UI platforms offer, also makes it easy for developers to take advantage of familiar APIs and idioms when porting games from other platforms (see Chapter 7 for details).

The availability of documentation (such as this book!), SDKs, and technical support for game developers is increasing, enabling them to design and implement high-quality games and get them to market faster. A good example for small developer teams is the free Forum Nokia remote device access (RDA) service, which enables testing on different phones to ensure consistency across the range of phones, without having to purchase them all.

For professional developers, the N-Gage platform goes further, by guaranteeing the game to run on N-Gage enabled devices, if written against the N-Gage SDK (see Chapter 8 for more details). Other middleware suppliers, such as Ideaworks3D, offer alternative solutions to compatibility and portability across a range of mobile devices (see the Airplay appendix at the end of this book).

In general, improvements in the network speeds available and flat-rate data charging, as described in the previous section, mean that larger game binaries can be downloaded, with better graphics, audio and video, as well as opportunities for multiplayer games and social game playing.

1.14 Summary

This chapter has raced through some of the aspects of mobile game development to give an overview of the industry for those wanting to create games for Symbian OS smartphones. It serves as a basic introduction to a huge topic, and gives some context, before diving into the technical details found in the rest of the book.

In this chapter, I've described the fact that, as a games platform, smartphones are different to consoles, PCs, and even portable game players like the Nintendo DS. There are challenges to writing games for a 'Swiss army knife,' but also opportunities to exploit the services available. You can write casual or hardcore games, or something in between. FlashLite is available on a number on Symbian smartphones in each global region. Java ME is available on all Symbian S60 and UIQ smartphones, and DoJa – described in Chapter 10 – is available in Japan on FOMA phones. For rich content games, native Symbian OS C++ APIs are provided in S60 and UIQ SDKs, and a range of middleware platforms and standards support is also available.

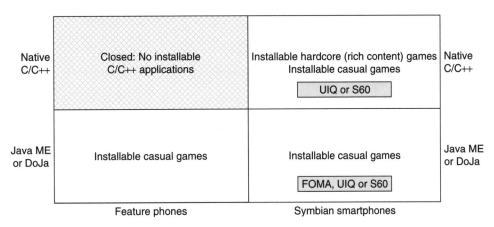

Figure 1.5   The types of installable games that can be created for feature phones and Symbian smartphones

Figure 1.5 reflects the types of game that can be created and installed on Symbian smartphones using Java ME and native C/C++, and contrasts the options available for creating installable games for feature phones.

The message of this chapter? Symbian smartphones sell in large volumes compared to portable game players. Symbian OS is set to be found in more mainstream phones than ever, because of the new pricing model and technology available for Symbian OS v9. It has never been a better time to create games for Symbian smartphones!

Part Two

Creating Native Games on Symbian OS v9

2

Symbian OS Game Basics

Jo Stichbury

2.1 Introduction

This chapter describes some of the basic issues to consider when creating a game on Symbian OS v9 in C++. We discuss how to write a standard game loop using an active object timer, how to handle user input from the keypad and screen, and discuss how to pause the game when interrupted by a change in application focus. Some of the resource limitations associated with the mobile platform (such as memory, disk space and floating point math support) are also discussed.

A sample game skeleton, imaginatively called **Skeleton**, which illustrates the points discussed, is available for download from **developer. symbian.com/gamesbook** and runs on both UIQ 3 and S60 3rd Edition emulators and hardware (it was tested on Sony Ericsson M600i, Motorola Z8, Nokia N73 and E61). The code uses the basic application framework for each UI platform generated using the New Project wizard available with Carbide.c++ v1.2. Very little additional UI-specific code has been added, because the basics described here use generic Symbian APIs. However, where differences exist between the platforms, they are described below.

2.2 The Game Loop

Most applications on Symbian OS are driven by user input, directly or indirectly. For example, the Calendar application displays text as a user types it, or responds to commands the user submits through the UI (such as formatting text, or creating and deleting appointments). If the user stops interacting with the application, the thread is suspended and only continues executing on receipt of, for example:

- an event generated when the user next does something, for example, by pressing a key or tapping the screen

- a system event such as a timer completion

- a notification of a change of focus, such as when the user switches to use another application, or the application is sent into the background by an incoming call or other event. For instance, the Calendar application may have an outstanding timer that expires when the application needs to display a notification dialog to the user to remind them of their next appointment.

The Symbian OS application model is said to be event-driven, which means that it does not run on a tight polling loop, constantly checking for input or changes of system state, but instead waits to be notified to run when it needs to respond to an event. This makes the threads that do need actually need to run more responsive, because they don't have to compete for a time slice. And when no threads need to run, Symbian OS can enter a power-saving state where all threads are suspended, thus optimizing the battery life.

Games, and other applications with constantly changing graphics, have different requirements to the event-driven applications described above, because they must continue to execute regularly in order to update their graphics and perform other calculations (such as detecting collisions or displaying countdown timers on their screens). For a game to execute regularly, regardless of the user input received, it is driven by what is known as a *game loop*, which is a loop that runs regularly for the duration of the game.

Typically, each time the game loop runs:

- The time elapsed since the last time the game loop executed is calculated, and any user input that has occurred since then is retrieved.

- The game engine updates the game internals, taking into account factors such as the time elapsed and the user input received.
  For example, in a game like **Asteroids**,[1] the user's key presses are used to calculate the new position of the player's ship, based on the actual input received and the time that has passed since the last graphics update. Besides making calculations to correspond to the player's input, the game engine also computes the new positions of all other moving graphics elements relative to the time of the last redraw. The

---

[1] See **developer.symbian.com/roidsgame** for a paper which describes writing an **Asteroids** clone for Symbian OS v9.

game engine must determine if there have been any collisions between the graphics elements and calculate their resulting status, such as their visibility (Have they moved off screen? Is one asteroid overlapping the others? Has a collision broken an asteroid into fragments?), and their changes in shape or color. The game engine also calculates status values, such as the player's score or the ammunition available.

- The screen is updated according to the new state calculated by the game engine.
  Having calculated the internal game state to reflect the new positions and status of each game object, the game engine prepares the next image to display on the screen to reflect those changes. Once the new frame is prepared, it is drawn to the screen. Each time the screen is updated, the new image displayed is known as a frame. The number of frames written to the screen per second is known as the frame rate. The higher the number of frames per second (fps), the smoother the graphics transitions appear, so a desirable frame rate is a high one in games with rapidly changing graphics. However, the exact figure depends on the game in question, affected by factors such as the amount of processing required to generate the next frame and the genre of game, since some games, such as turn-based games, do not require a high frame rate.

- The game's audio (background music and sound effects) are synchronized with the game state.
  For example, the sound of an explosion is played if the latest game update calculated that a collision occurred.

Figure 2.1 shows this graphically, while in pseudocode, a basic game loop can be expressed as follows:

```
while ( game is running )
  {
  calculate time elapsed and retrieve user input
  update game internals
  update graphics (prepare frame, draw frame)
  synchronize sound
  }
```

Note that a constantly running game loop is not necessary in every game. While it is needed for a game with frequently changing graphics in order to update the display, some games are more like typical event-driven applications. A turn-based game, like chess, may not need a game loop to run all the time, but can be driven by user input. The loop starts running when a turn is taken, to update the graphics to reflect the move, and calculate the AI's response, then returns to waiting on an event to indicate the user's next move or other input, such as deciding to quit the game.

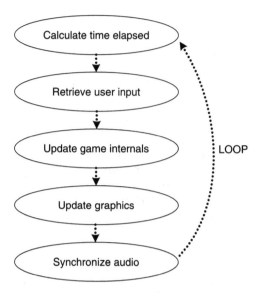

Figure 2.1 A graphical representation of a game loop

2.3 The Heartbeat Timer

On Symbian OS, it is important not to hog the CPU so you should not run a tight synchronous loop like the one shown in pseudocode above. The game loop must return control to the thread's active scheduler to allow other system events to run as necessary, allowing other threads in the system the opportunity to run. If the game loop prevents the active scheduler from running a particular active object associated with the view server regularly, between 10 and 20 seconds, the infamous EVwsView-EventTimeOut panic occurs. This terminates the game thread with panic category ViewSrv and error code 11.

So instead of trying to work around this, on Symbian OS, a game loop should be run using a timer active object, such as CTimer or CPeriodic, to trigger the game loop to run regularly. It is better to use a single timer to run one iteration of the game loop than to use multiple timers to control different time-based events in the game. The latter approach can lead to very complex code, since there may be many elements in a game that respond after different timeouts have expired, which would be difficult to synchronize and to debug. Active objects cannot be pre-empted, so the accuracy of a timer active object is also dependent on other active objects in the system. If multiple timers are used for different events, the combination could lead to significant timer drift for some lower-priority active objects. Additionally, each timer consumes kernel resources – if a large number are required, this could add up to an undesirable system overhead.

The solution is to use a single timer object to provide a game 'heartbeat.' All elements of the game that require a particular time-based activity use the heartbeat to synchronize themselves. The timer fires as many times per second as necessary to achieve the desired frame rate. Each time the timer fires, it runs the game loop once, and increments a stored value known as the tick count. The tick count is used to provide the time elapsed to the rest of the game.

The resulting pseudocode for the game loop looks as follows, where `TimerEventHandler()` is the method called once every heartbeat to handle the timer event:

```
TimerEventHandler()
  {
  calculate the time elapsed and retrieve user input
  update game internals
  update graphics (prepare frame, draw frame)
  synchronize sound
  }
```

One of the benefits of using a single heartbeat timer is that the beat rate can easily be slowed down for debugging purposes, stopped entirely when the game needs to pause, or tweaked to modify the frame rate when the graphics performance differs across a range of devices with different capabilities.

The `CPeriodic` timer is used most commonly to drive a game loop. It uses a callback function to execute the game loop on each heartbeat. `CPeriodic` is the simplest class to use to provide a timer without coming into direct contact with active objects because it hides the details of the active object framework (it derives from `CTimer` – an active object-based timer class – and implements the active object methods `RunL()` and `DoCancel()` for its callers). The **Skeleton** example uses this class, as shown below. However, if you already use a number of active objects in your game, you may simply prefer to use the `CTimer` class directly.

```
void CSkeletonAppView::ConstructL( const TRect& aRect )
  {
  ...
  // Create the heartbeat timer for the game loop.
  iPeriodicTimer = CPeriodic::NewL(CActive::EPriorityStandard);
  // Activate the window (calls StartHeartbeat())
  ActivateL();
  }

void CSkeletonAppView::StartHeartbeat()
  {
  // Start the CPeriodic timer, passing the Tick() method for callback
  if (!iPeriodicTimer->IsActive())
    iPeriodicTimer->Start(KFramesPerSecond, KFramesPerSecond,
                TCallBack(CSkeletonAppView::Tick, this));
  }
```

```
// Tick() callback method
TInt CSkeletonAppView::Tick(TAny* aCallback)
  {
  ASSERT(aCallback);
  static_cast<CSkeletonAppView*>(aCallback)->GameLoop();
  // Return a value of ETrue to continue looping
  return ETrue;
  }

void CSkeletonAppView::GameLoop()
  {
  //  Update the game internals - details omitted here
  }
```

As I mentioned above, because Symbian OS active objects cannot be pre-empted, if another active object event is being handled when the timer expires, the timer callback is blocked until the other event handler completes. The timer's event processing is delayed, and the timer request is not rescheduled until it runs. This means that periodic timers are susceptible to drift, and will give a frame rate slightly lower than expected. Additionally on Symbian OS v9, the standard timer resolution is 1/64 seconds (15.625 milliseconds) on both hardware and the emulator. So, for example, if you request a period of 20 milliseconds, you actually receive an event approximately once every 30 milliseconds (that is, 2 × 15.625 milliseconds). What's more, if you've just missed a 1/64th 'tick' when you submitted the timer, then it'll only be submitted on the next tick, which means that the first period will be almost 3 × 15.625 (=48.875) milliseconds. This can add up to a significant jitter.

As an example, in the **Skeleton** example, I set the timer period to be 1/30 second (33.3 milliseconds), which at first sight could be expected to deliver 30 fps The frame rate actually observed was relatively constant at 22 fps occasionally dropping to 21 fps. To deliver 30 fps, I would need to increase the period and add in logic to see how much time has passed between each tick, and then compensate for it when rescheduling the timer.

Symbian OS does actually provide a heartbeat timer class, called CHeartBeat to do this. This class is similar to the periodic timer, except that it provides a function to restore timer accuracy if it gets out of synchronization with the system clock. The CHeartBeat class accommodates the delay by calling separate event-handling methods depending on whether it ran accurately, that is, on time, or whether it was delayed, to allow it to re-synchronize. However, this additional functionality is not usually necessary. Extremely accurate timing isn't generally required because a game engine only needs to know accurately how much time has elapsed since the previous frame, so, for example, the total amount of movement of a graphics object can be calculated.

There is also a high resolution timer available through the CTimer class, using the HighRes() method and specifying the interval required in microseconds. The resolution of this timer is 1 ms on phone hardware, but defaults to 5 ms on the Windows emulator, although it can be changed by modifying the `TimerResolution` variable in the `epoc.ini` emulator file. However, on Windows, there is no real-time guarantee as there is on the smartphone hardware, because the timings are subject to whatever else is running on Windows.

2.4 Handling Input from the Keypad

Handling user input made by pressing keys on the phone's keypad is quite straightforward in the Symbian OS application framework. The application UI class should call `CCoeAppUI::AddToStackL()` when it creates an application view to ensure that keypad input is passed automatically to the current view. The key events can be inspected by overriding `CCoeControl::OfferKeyEventL()`.

Note that only input from the numerical keypad and multi-way controller is passed into `CCoeControl::OfferKeyEventL()`. Input from the user's interaction with the menu bar, toolbar, or softkeys, is passed separately to the `HandleCommandL()` method of the application UI class for S60. In UIQ, the application view handles in-command input. This is because a command can be located on a softkey, in a toolbar, or in a menu, depending on the interaction style of the phone. To allow for this flexibility, commands are defined in a more abstract way, rather than being explicitly coded as particular GUI elements, such as menu items.

For either UI platform, the input event is handled according to the associated command, which is specified in the application's resource file. The **Skeleton** example demonstrates basic menu input handling in `CSkeletonAppUi::HandleCommandL()` for S60 and `CSkeletonUIQView::HandleCommandL()` for UIQ.

In the **Skeleton** example, the handling of the user input is decoupled from the actual event, by storing it in a bitmask (`iKeyState`) when it is received, and handling it when the game loop next runs by inspecting `iKeyState`. This approach is shown in the simplified example code, for S60, below.

```
TKeyResponse CSkeletonAppView::OfferKeyEventL(const TKeyEvent&
                                  aKeyEvent, TEventCode aType)
  {
  TKeyResponse response = EKeyWasConsumed;
  TUint input = 0x00000000;

  switch (aKeyEvent.iScanCode)
    {
    case EStdKeyUpArrow: // 4-way controller up
      input = KControllerUp; // 0x00000001;
      break;
    case EStdKeyDownArrow: // 4-way controller down
```

```
      input = KControllerDown; // 0x00000002;
      break;
   case EStdKeyLeftArrow: // 4-way controller left
      input = KControllerLeft; // 0x00000004;
      break;
   case EStdKeyRightArrow: // 4-way controller right
      input = KControllerRight; // 0x00000008;
      break;
   case EStdKeyDevice3: // 4-way controller center
      input = KControllerCentre;
      // This is the "fire ammo" key
      // Store the event, the loop will handle and clear it
      if (EEventKey == aType)
         {
         iFired = ETrue;
         }
      ...
   default:
      response = EKeyWasNotConsumed;// Keypress was not handled
      break;
   }

if (EEventKeyDown == aType)
   {// Store input to handle in the next update
   iKeyState | = input; // set bitmask
   }
else if (EEventKeyUp == aType)
   {// Clear the input
   iKeyState &= ~input; // clear bitmask
   }

return (response);
}
```

TKeyEvent and TEventCode are passed in as parameters to Offer-
KeyEventL(), and contain three important values for the key press – the
event type, scan code, and character code for the key. Let's discuss these
in some further detail.

2.4.1 Key Events

Key events are delivered to the application by the Symbian OS window
server (WSERV). WSERV uses a typical GUI paradigm which works well
for controls such as text editors and list boxes. That is, when a key is
held down, it generates an initial key event, followed by a delay and
then a number of repeat key events, depending on the keypad repeat rate
setting, until it is released.

There are three possible key event types. Each key press generates the
three event types in sequence:

1. EEventKeyDown – generated when the key is pressed down

2. `EEventKey` – generated after the key down event, and then regularly as long as the key is kept down[2]

3. `EEventKeyUp` – generated when the key is released.

The `EEventKeyDown` and `EEventKeyUp` pair of events can be used to set a flag in the game to indicate that a particular key is being held, and then clear it when it is released, as shown in the previous example. Using this technique, the state of the relevant input keys is stored, and key repeat information can be ignored (the repeated input notification often does not work well for games, because it results in jerky motion, for example, when a user holds down a key to drive a car forward in one direction). However, it can be useful in other contexts. Since the `EEventKey` event is passed repeatedly if a key is held down, it can be used for repeating events, such as firing ammunition every quarter second while a particular key is held down. This is much easier and more comfortable for the player than needing to repeatedly press the key. The ***Skeleton*** example code, shown above, illustrates this each time a repeat event is received from holding down the centre key of the 4-way controller.

2.4.2 Key Codes

Either the scan code or the character code can be used by the game to determine which key has been pressed. The scan code (`TKeyEvent::iScanCode`) simply provides information about the physical key pressed, which is usually sufficient for processing events in games, since all that is required is information about the direction or the keypad number pressed. The character code may change when, for example, a single key is pressed multiple times, depending on how the mappings and any front end processors are set up. This level of detail isn't often needed by a game, so the character code is generally ignored.

A game must also be able to handle the case where two or more keys are pressed simultaneously, for example, to simulate 8-way directional control or allow for more complicated key sequences and controls. The ***Skeleton*** example does this by storing input in a bitmask, so multiple simultaneous key presses can be handled together by the loop.

By default, pressing multiple keys simultaneously is not possible; the event from the first key to be pressed is received and others are discarded while it is held down. This is called key blocking. However, the application UI can call `CAknAppUi::SetKeyBlockMode()` on

---

[2] `RWsSession::GetKeyboardRepeatRate()` can be used to determine the repeat interval and it can be modified by calling `RWsSession::SetKeyboardRepeatRate()`.

S60 to disable key blocking for the application and accept multiple key press events. However, this is not possible on UIQ.

One other thing to note about input is that phone keypads have a limited number of keys, which are not always ergonomic for game play. The variety of layouts and the size limitations need to be considered when designing a game, and ideally should be configurable, through a menu option, so the user can choose their own, according to their preference and the smartphone model in question. It is also a good idea to design games to always allow for one-handed play, since some devices are too small for game playing to be comfortable using both hands. However, other devices are suited to two-handed game playing, particularly when they can be held in landscape orientation, such as the Nokia N95 or the Nokia N81. The instance where phones can be used in both portrait and landscape orientations, and the game supports each mode, is another good reason for allowing the game keys to be configurable. The user can set different key configurations for gameplay depending on the orientation of the phone.

2.5   Handling Input from the Screen

Some UIQ smartphones, such as the Sony Ericsson P1i, support touch-screen input, which can be used to add another mode of user interaction to a game. Screen events can be detected and handled in much the same way as keypad input, by overriding `CCoeControl::HandlePointer-EventL()` as shown below. In the example given, to capture events that occur in the close vicinity of a particular point on the screen, an area of the screen is constructed, ten pixels square, and centered on the point in question. When a pointer-down screen event occurs, if the point at which the tap occurred lies within the square, the user input is stored in a bitmask in the same way as for a key press.

```
void CSkeletonUIQView::HandlePointerEventL(const TPointerEvent&
                                           aPointerEvent)
  {
  if (TPointerEvent::EButton1Up == aPointerEvent.iType)
    {// Pointer up events clear the screen
    iKeyState = 0x00000000; // clear all previous selections
    }
  else if (TPointerEvent::EButton1Down == aPointerEvent.iType)
    {
    TRect drawRect( Rect());
    TInt width = drawRect.Width();
    TInt height = drawRect.Height();

    TPoint offset(10,10); // 10x10 square around the screen position

    TPoint k1(width/4, height/2);
```

```
  TRect rect1(TPoint(k1-offset), TPoint(k1+offset));
  if (rect1.Contains(aPointerEvent.iPosition))
    {
    iKeyState | = KKey1;
    return; // stored the event, so return
    }

  TPoint k2(width/2, height/2);
  TRect rect2(TPoint(k2-offset), TPoint(k2+offset));
  if (rect2.Contains(aPointerEvent.iPosition))
    {
    iKeyState | = KKey2;
    return; // stored the event, so return
    }
  ... // Other numbers similarly inspected
  }

// Pointer events for other areas of the screen are ignored
// Pass them to the base class
CCoeControl::HandlePointerEventL(aPointerEvent);
}
```

Figure 2.2 illustrates the **_Skeleton_** example in the UIQ emulator. The mouse is clicked and held on, or near, the '8' digit to simulate a tap and hold on the screen. The event is detected through `CSkeletonUIQView::HandlePointerEventL()` as shown above for digits '1' and '2.' The `iKeyState` member variable is updated to reflect the screen event, and next time the game loop runs and the screen is updated, the display for that digit highlights that the stylus is held to the screen. When the stylus (or mouse click, in the case of the emulator) is released, a pointer-up event is received and the highlight is removed.

Figure 2.2 Handling a pointer event received from a screen tap

2.6   System Events

As we've discussed, the game loop is driven by a timer active object and runs regularly to update the game engine internals, resulting in a given number of frames per second. For any length of time, on a mobile operating system, this kind of regular looping can lead to a drain in battery power. The requirement to regularly run the game thread prevents the OS from powering down all but the most essential resources to make efficient use of the battery.

2.6.1   Loss of Focus

As a developer, you need to be aware that the game is consuming battery power whenever it is running the loop, and ensure that the heartbeat timer runs only when required. For example, the game loop should be paused when the game is interrupted by a system event, such as an incoming call, or because the user has tasked to another application. The game loop should also pause if the user stops interacting with the game for any length of time; for example, if she is playing the game on the bus and has to get off at her stop without quitting the game before doing so.

You should implement `CCoeControl::FocusChanged()` to be notified of a change in focus of the game – that is, when another application or system dialog sends the game to the background. As the example shows, the timer is halted when the focus is lost, and started when it is regained. The benefit of having a single heartbeat time to drive the game loop is that it is very straightforward to stop and start it when necessary.

```
void CSkeletonAppView::FocusChanged(TDrawNow aDrawNow)
  {
  if (IsFocused())
    {// Focus gained
    StartHeartbeat(); // Starts the gameloop
    }
  else
    {// Focus lost
    if (iPeriodicTimer->IsActive())
      {
      StopHeartbeat(); // Pauses the game loop
      }
    // Save game data here if necessary
    }

  if(aDrawNow)
    DrawNow();
  }
```

2.6.2   User Inactivity

If no user input is received, after a certain length of time, to minimize the phone's power usage, the system will gradually dim the backlight, and eventually turn it off completely. The length of inactivity time before each step occurs is often customizable by the phone's owner, and usually, a screensaver is displayed, such as the time and date, which can also be customized.

Regardless of the fact that the phone appears to be in power-saving mode, the game loop will continue running, even when the screensaver is showing and the backlight is off, unless the developer adds code to explicitly stop it. To save power, it is usual practice for the game to move into 'attract' mode, or to display a pause menu and stop the game loop after user inactivity of a certain duration.

Attract mode is so called because it displays a graphics sequence which looks something like the game running in a demo mode, showing off the game to attract the user to play it. Attract mode also consumes power, because it is still displaying regularly updated graphics and running a game loop, even if the frequency of the timer is much reduced from that of normal gameplay. Unless the phone is running on main power (i.e., connected to its charger), it is usual to run the attract mode for a pre-determined amount of time only, perhaps two minutes, and then display the pause menu and halt the game loop by completely stopping the heartbeat timer.

User inactivity can be detected by creating a class that inherits from the `CTimer` active object and calling the `Inactivity()` method. The object receives an event if the interval specified, passed as a parameter to the method, elapses without user activity.

```
class CInactivityTimer : public CTimer
    {
public:
    static CInactivityTimer* NewL(CSkeletonAppView& aAppView);
    void StartInactivityMonitor(TTimeIntervalSeconds aSeconds);
protected:
    CInactivityTimer(CSkeletonAppView& aAppView);
    virtual void RunL();
private:
    CSkeletonAppView& iAppView;
    TTimeIntervalSeconds iTimeout;
    };

CInactivityTimer::CInactivityTimer(CSkeletonAppView& aAppView)
: CTimer(EPriorityLow), iAppView(aAppView)
    {
    CActiveScheduler::Add(this);
    }

CInactivityTimer* CInactivityTimer::NewL(CSkeletonAppView& aAppView)
    {
    CInactivityTimer* me = new (ELeave) CInactivityTimer(aAppView);
    CleanupStack::PushL(me);
    me->ConstructL();
    CleanupStack::Pop(me);
    return (me);
    }
```

```
void CInactivityTimer::StartInactivityMonitor(TTimeIntervalSeconds
                                                          aSeconds)
  {
  if (!IsActive())
    {
    iTimeout = aSeconds;
    Inactivity(iTimeout);
    }
  }

void CInactivityTimer::RunL()
  {// Game has been inactive - no user input for iTimeout seconds
  iAppView.SetPausedDisplay(ETrue); // Displays the pause screen
  iAppView.StopHeartbeat(); // Stop the game loop
  }
```

The `CInactivityTimer::StartInactivityMonitor()` meth-od must be called each time the game loop starts, to monitor for user inactivity while it is running.

Following any kind of pause, when the user returns to the game and it regains focus, the game should remain in the paused state until the user makes an explicit choice to proceed, quit, or perform another action, depending on what the pause menu offers.

The usability of the pause menu is a hot topic – it's been said that you can tell the quality of a mobile game from how easy it is to resume from the pause menu. This aspect of game design is outside the scope of this book, but you can find an interesting paper called *At The Core Of Mobile Game Usability: The Pause Menu* in the Usability section of the Forum Nokia website (***www.forum.nokia.com/main/resources/documentation/usability***).[3] This section has a number of other useful documents for mobile game developers and designers, and I recommend that you browse them regularly, even if you are developing a mobile game for the other Symbian OS UI platforms.

2.6.3 Simulating User Activity

There are occasions when the game may want to simulate activity. For example, while attract mode is playing, it is desirable to prevent the system from dimming the backlight, the screensaver taking over, and eventually the backlight turning off completely. This may seem contradictory, but often games will detect user inactivity, move to attract mode, and then actually simulate user activity while attract mode is running, so the graphics can be seen. User activity is simulated by calling

---

[3] The exact location of the paper has a URL which is far too long to type in from a book, and is subject to change anyway, if the content is updated in future. We will keep a set of links to papers and resources that are useful to game developers on the Symbian Developer Network Wiki (***developer.symbian.com/wiki/display/academy/Games+on+Symbian+OS***).

`User::ResetInactivityTime()` regularly, in the game loop. This is unnecessary while the game is being played, because the user is generating input. It should also cease when the game changes modes to display the pause menu, to then allow the system to switch off the backlight and display the screensaver as it normally would do.

Since simulating user activity during attract mode prevents the system from saving power by dimming its backlight, the length of time the game displays its attract mode should be limited, as section 2.6.2 described, unless the phone is powered externally by the main charger. The `CTelephony` class can be used to retrieve information about whether the phone is connected to a main charger, by calling `CTelephony::GetIndicator()`. To use `CTelephony` methods, you must link against `Etel3rdParty.lib`, and include the `ETel3rdParty.h` header file.

2.6.4 Saving Game Data

When the game is paused, it should save any important game state in case it is needed to continue the game in the state it was in before pausing. For example, the player may re-boot the phone before the game is next played, but it should still be able to continue the game from the point before it was paused. A player who has got to a particular level in a game will not be happy if their progress is lost because they had to interrupt their session.

A game should always save its data at regular intervals so it can be restored to the same state, should the game suddenly be stopped, for example if the battery power fails or the battery is removed completely. In the event of a 'battery pull,' the game's data manager code should be prepared for the possibility that the data has been corrupted – such as if an update was under way when the power down occurred. The game should be able to recover from finding itself with corrupt data, for example, by having a default set available in replacement, or performing data updates using rollback.

When saving game data, it is important to consider the amount of information that should be written to file. If there is a large amount of data, for example over 2 KB, it is advisable not to use the blocking `RFile::Write()` method, since this could take some time to complete and would block the game thread. It is advisable to use an active object to schedule the update, using the asynchronous overload of `RFile::Write()`, or to use one of the higher-level stream store APIs.

2.7 Memory Management and Disk Space

Mobile devices have limitations on the amount of memory (RAM) they can support for a number of reasons, including size (the limitations of the physical space available), cost, and current draw (the amount of power required to access the memory). Symbian OS is a multitasking operating system, meaning that other applications are running alongside the game. So, because the memory resources available are finite, the available memory may suddenly be very limited.

It would make a bad user experience if, halfway through shooting an enemy, the game crashed or closed because there was insufficient memory to continue. It is important, therefore, for a game to reserve all the memory it will need at start up, so it can guarantee to satisfy all memory allocation requests. If the memory cannot be reserved, the game startup should fail and ask the user to shut down other applications before trying to play.

It is quite straightforward to allocate a pool of memory to a game, by setting the *minimum* size of its heap using the EPOCHEAPSIZE specifier in its MMP file.[4] If the game process succeeds in launching, then this amount of memory is available to the heap – for all memory allocations in the main thread, or any other thread within the process, as long as it uses the same heap. This approach makes writing code that allocates heap-based objects within the game quite straightforward – you don't have to check that the allocation has succeeded in release builds, as long as you profile and test every code path to ensure that, internally, the game does not use more memory than was pre-allocated to it.

However, if your game launches any separate processes (e.g., Symbian OS servers) or places heavy demands on those already running (for example, by creating multiple sessions with servers, each of which uses resources in the kernel) then this approach will not be sufficient to guarantee the game will run if the phone's free memory runs low. Again, the game should attempt to allocate the resources it needs, including those in other processes, at startup – or have a means of graceful failure if this is not practical and a later resource allocation fails.

When pre-allocating memory for the game, it is important to make an educated guess at what is required, rather than simply reserve as much of the memory as is available on the device. Games for the N-Gage platform, for example, which are typically the highest quality rich content games available on Symbian OS v9, limit themselves to a maximum allocatable heap size of 10 MB.[5] So if you are writing a basic casual game, you should certainly try to stay well under this limit! Your game will not be popular if it prevents normal use of the phone, such that it

---

[4] The minimum size of the heap specifies the RAM that is initially mapped for the heap's use. By default it is set at 4 KB. The process can then obtain more heap memory on demand until the maximum value, also set through use of the EPOCHEAPSIZE specifier, is reached. By default, the maximum heap size is set at 1 MB.

[5] N-Gage games are tested and certified prior to release (as we'll describe in Chapter 8) and the 10 MB limit is enforced throughout the launch, execution and termination of the game.

cannot run in parallel with other applications. Since many applications for Symbian OS v9 require Symbian Signing, which verifies the general behaviour and 'good citizenship' of the application, you need to stay within self imposed limits. You should also consider advice and best practice for memory usage on Symbian OS, including regular testing for memory leaks, for example, by using debug heap checking macros like __UHEAP_MARK. Wherever possible, de-allocate and reuse memory that is no longer required to minimize the total amount your game needs.

Some developers create memory pools for managing their memory allocations in order to minimize fragmentation of the heap, use memory more efficiently, and often improve performance. A common approach is to create a fixed-size pool for persistent data and then use a separate pool for transient data, which is recycled when it is no longer needed in order to minimize fragmentation.

You can find more information about working efficiently with limited memory in most Symbian Press C++ programming books, and general advice in *Small Memory Software: Patterns for Systems with Limited Memory* by Noble and Weir, listed in the References chapter at the end of this book. There is also a set of tools available for checking memory usage and profiling code on Symbian OS. You can find more information about these from the Symbian Developer Network (***developer.symbian.com***).

Disk space refers to the non-volatile shared memory used to store an installed game, rather than the main memory used to store the game's modifiable data as it executes. Some Symbian smartphones, like the Nokia N81 8 GB and N95 8 GB have large amounts of space (8 GB) for internal data storage, but the typical internal storage size for other Symbian OS v9 smartphones can be as limited as 10 MB. However, users can extend their storage with removable drives such as mini SD or micro SD cards or Memory Sticks. The drive letter for the removable media may vary on different handsets and UI platforms (for example, the removable media is sometimes the D: drive and sometimes the E: drive). You can use the `RFs::Drive()` method to determine the type of drive, and `RFs::Volume()` to discover the free space available on it.

When installing a game, there must be sufficient disk space to contain the installation package for the game, and subsequently decompress and install the game and its asset files. It is desirable to keep a game footprint – the storage space it occupies – as small as possible if it is to be downloaded over-the-air (OTA), since a very large game will be more costly and slower to download over a phone network.

2.8 Maths and Floating Point Support

Games that require a number of physics-heavy calculations (for example, those modeling acceleration, rotations, and scaling) typically use floating point numbers. Most current ARM architectures do not support floating point instructions in hardware, and these have to be emulated in software.[6] This makes the use of floating point numbers expensive in terms of processing speed when thousands of operations are required. Where possible, calculations should instead be implemented using integers or fixed-point arithmetic, or they should be performed in advance on a PC and stored in static lookup tables to be used at run time in the game. It is rare that high precision is required for geometric operations such as sin, cos, and square root, so there is room for optimization, for example, for sin and cos, the Taylor series can be used to approximate sin and cos to a number of decimal places.

The use of 32-bit integers should be preferred where possible; 64-bit integers have run-time performance and memory consumption overheads, and are rarely needed in practice.

2.9   Tools and Further Reading

This chapter has discussed some of the basic aspects of creating a well-behaved game loop, handling input, managing interruptions and user inactivity, and coping with resource limitations.

Where the performance of the game is critical, we advise you to analyze the code for each device it is to run on, since devices and platforms can have very different characteristics. A good tool for performance analysis is the Performance Investigator, delivered with Carbide.c++ Professional and OEM Editions. A profiling tool, it runs on an S60 3rd Edition device and collects information about processes, memory, CPU usage, and battery usage. Within Carbide.c++, the trace data collected can be reviewed and analyzed. More information, and a white paper that describes the Performance Investigator tool, giving an example of how to use it, can be found at *www.forum.nokia.com/carbide*.

For more information about general C++ development on Symbian OS, we recommend the Symbian Press programming series, described in the References and Resources section of this book. There are also a number of papers available on the developer websites listed in that section, including a discussion forum specifically for C++ games development for S60 on Forum Nokia at *discussion.forum.nokia.com/forum*.

---

[6] Some of the high-end phones based on ARM11 processors do have floating point support, for example, the Nokia N95. To use it, code must be compiled with a particular version of the RVCT compiler, with floating point hardware support options enabled. The resultant binaries do not execute on hardware without floating point support.

3

Graphics on Symbian OS

Twm Davies
(twmdesign.co.uk/theblog)

3.1 Introduction

This chapter aims to provide the reader with an understanding of the fundamental graphics services on Symbian OS. It covers a range of topics from the basics of drawing geometric shapes to the playback of video. At the same time, it addresses some of the challenges of drawing efficient graphics to the screen.

Symbian OS is a generic mobile OS and, as such, provides APIs to suit a variety of graphical applications, from list boxes and rich text editors to complex 3D projections. As a result, the range of graphics APIs is vast and there seems to be many ways of achieving the same task. This chapter focuses on a cross section of APIs and usage patterns which are most useful to have in a game developer's tool chest and provides fragments of code which demonstrate correct usage of Symbian APIs.

The discussion and example code projects found within this chapter are based on Symbian OS v9.2, tested on the Nokia N95 and Sony Ericsson M600i (based on Symbian OS v9.1) smartphones. I expect the content to remain compatible with changes to the Symbian OS graphics architecture that are anticipated in v9.5.

Speaking of which, in October 2007, Symbian announced ScreenPlay (*www.symbian.com/symbianos/screenplay*) which is the name for the new graphics architecture for Symbian OS. It provides a window composition engine much like that found on Apple OS X and Microsoft Vista. At the time of writing, there isn't much additional information available, but by the time this book is published you can expect to see more information about ScreenPlay on the main Symbian website at *www.symbian.com*.

3.1.1   Example Code

This chapter makes occasional reference to the **Roids** example game on the Symbian Developer Network website (***developer.symbian.com***) which is a complete game with full code made available for UIQ 3 and S60 3rd Edition. Just search for 'Roids' at the main page, or navigate directly to ***developer.symbian.com/ roidsgame***.

Code fragments beginning with CGfx are part of the example code to go with this chapter. The example code covers the following:

- drawing to the screen using a graphics context (GC)

- directly manipulating the pixels of the frame buffer

- synchronizing updates using the anti-tearing API

- loading images using the image conversion library (ICL)

- video playback using the Symbian video playback utility.

The example code can be downloaded from the Symbian Press website ***developer.symbian.com/main/learning/press***, and from the book's page on the Symbian Developer Network wiki (***developer.symbian.com/wiki***).

3.2   Overview

The original graphic services provided by Symbian OS evolved from the need to supply a PDA UI toolkit which shared similarities with desktop systems. The windowing system and application model were designed with multitasking in mind and have always been focused on effective sharing of the display.

Though the early frameworks of Symbian OS are very much alive in Symbian OS v9.x, a whole range of additional APIs and features have been developed to power the visually rich applications and games built for Symbian smartphones of late. Modern Symbian devices are expected to deliver sophisticated 3D visuals, smooth animation and bi-directional video streaming out of the box, which is good news for game developers since many of these APIs are part of the public SDKs for Symbian C++ developers.

Not all devices are equal; the role of an operating system, and any middleware running on top of it, is to shield applications from differences in hardware. However, games have traditionally been tuned and released separately for specific devices. It's very challenging to create a pleasing cross platform game, since gameplay and performance can be influenced by the differences in basic hardware characteristics, particularly the screen and input methods of a device.

Symbian OS devices offer little relief to those demanding a homogenous platform since handset capabilities, software versions, and screen sizes differ, depending on the market segment and the device manufacturer.

3.2.1 Cross Platform Support

To illustrate some of the problems with supporting multiple devices, consider the deployment of a simple very game. **Roids** is a simple **Asteroids** clone which was originally written and tested on S60-based devices and the UIQ emulator.[1] **Roids** relies on a 4-way cursor/joystick to give the user dexterous control of a spacecraft. When the UIQ test device arrived (a Sony Ericsson M600i), the SIS installation file for the game installed okay and the game launched; but there was no way of generating 4-way cursor events, since the M600i has no joystick!

A modification was required to simulate the joystick events by pressing other keys, which was functional but lost some of the feel of the version running on Nokia E61. The M600i does however have a touch screen and stylus, which makes it great for games optimized for touch.

To this end, it is sensible to set the expectation that, although S60 and UIQ have a strong binary compatibility within their respective platforms (for example S60 3.1 apps should work on S60 3.2 and onwards), some styles of games may require significant modifications to retain the usability or fluidity which characterizes the game, when run on devices different from that for which it was originally designed. A good tip to remember is that, before committing to using a particular API on a device, it's best to assess how widely it's supported and what the performance is like on the target device or devices.

A good level of device independence can be achieved by using graphics APIs with wide industry support such as OpenGL ES and OpenVG. These standards work particularly well for games, since 3D and vector game worlds can scale across screen sizes, allowing the developer to focus on the game design while the hardware engineers and handset manufacturers can work on optimizing the implementations to the hardware in question.

The reality is that a consistent user experience across a range of devices can only really be assured by either extensive testing or by establishing a minimum supported device configuration. Developer platforms such as the N-Gage platform, described in Chapter 8, use both approaches to provide a common target for game developers.

3.2.2 Hitting the Metal

Game developers have a long heritage of innovation in creating impressive smooth visuals and extracting optimum performance from a target

---

[1] As I mentioned in the introduction, you can find the code for **Roids**, and a white paper describing it, on the Symbian Developer Network at **developer.symbian.com/roidsgame**.

device. The term 'hitting the metal' refers to accessing a hardware resource, such as the screen or keyboard, directly from application code, that is, by avoiding the APIs the operating system wraps over it, dealing with interrupts, and setting up the hardware directly. This approach was possible when games were developed for dedicated game console hardware or the game was running as the sole task on a home computer.

As an example, an established technique on 1980s devices was to drive the game from the 'vertical blank.' This technique allowed program code to run in the periods when the display hardware was turned off, waiting for the screen to complete its vertical blank (which took about 20 µs). Since the vertical blank period occurred regularly every 25th or 30th of a second (depending on the television standard), timing the code to run during this period guaranteed a regular 'heartbeat' to help drive the game – which made synchronizing updates to the display trivial.

The general approach to taking over the machine was very wasteful in many ways, with game developers having to write and maintain mundane routines, such as reading from the disk, which made porting the game to another platform quite laborious. But despite this, the complete control offered by these environments meant that each CPU instruction was executed because the game developer said so, and games didn't need to worry about being pre-empted by phone calls, anti-virus software, and so on. Although the CPUs of those devices were hugely underpowered compared to today, games were essentially well bounded real-time systems which could meet the deadlines of high frame rates. Many of the game world archetypes and distinctive graphical effects evolved due to tricks discovered when using hardware and APIs not designed for games.

The challenges for mobile games are somewhat different. On a multi-tasking system, the applications are virtualized with direct access to the hardware hidden behind a kernel privilege boundary. It's common to have a user side component which exists solely to resolve contention of a scarce resource between multiple clients. Even with high CPU speeds and support from specialized graphical processing units (GPUs), the challenge of rendering immersive worlds and sustaining good performance is still a creative art. While a 'write once, run many' approach is best for deployment of games to the widest audience and hence makes the best return on an investment, history shows that specific innovations and compelling games may occur while developing for one particular device. This chapter, in places, refers to some specific APIs which may only be available on certain devices.

3.3   On Frame Rate

As Jo described in the previous chapter, the number of frames which are rendered to the screen each second is known as the frame rate, which is measured in frames per second (abbreviated to fps). The unit of frequency, Hertz (Hz), may also be used in relation to fps, but in general fps denotes a value which may change depending on the complexity of a game scene (for example, very complex explosions may result in a reduced frame rate), while frequency is used to represent the unchanging periodic update of the screen by the display hardware.

3.3.1 How Many Frames Per Second Are Enough?

In film and animation, fps defines the fixed rate at which images are captured or displayed to represent motion. The illusion of movement is created when still images are displayed so rapidly as to satisfy the viewer's visual interpretation of a continuous scene. For certain types of games, the frame rate and stability of the rate are very important for creating believable, immersive worlds. Within some parts of the PC games community, a culture of competition has emerged whereby developers build new systems and use techniques, such as over clocking and water cooling, to increase the processing speed and thus the frame rate of the game. When the rate approaches around 75 fps, the frequency reaches what is known as the flicker fusion factor, and the human eye cannot really distinguish higher frequencies.

Motion pictures in the cinema are usually played at 24 fps and television is broadcast at no more than 30 fps (in the case of the American NTSC broadcast standard). This would seem to indicate a reasonable target range of between 24–30 fps to aim for achieving credible motion, but that's not the whole story.

A frame of cinema film is not like a discrete animation frame. There is no such thing as an instance in film. A film camera has a rotary shutter which interrupts the light two or three times for every film frame, and so it exposes each frame to two or three steps of motion. Also, just like still camera film, cinema reel film is sensitive to all the light and hence captures the continuous motion which occurs during an exposure. The end result is that a lot of motion is compressed it into a single image. This is why even a DVD shows signs of motion blur when a frame is paused on an action scene. An analogy from the animation industry helps to illustrate this, as the following section will show.

3.3.2 Stop Motion Animation

Creating stop motion animation, such as Aardman Animation's 'Wallace and Gromit' films, is a laborious task. The film is played back at 24 fps but may be shot 'on twos' which means taking two exposures for each pose. This results in an effective 12 frames of animation per second when played back on a standard cinema film projector.[2]

---

[2] For more about the Wallace and Gromit stop motion rendering technique, see, for example, the Wikipedia entry at *en.wikipedia.org/wiki/Wallace_and_Gromit*.

At 12 fps, an animator wishing to move a character's head from left to right over a two second interval would need to issue 24 small increments to the character's head in order to get the desired timing and then issue two shots per motion so that it will play back at the correct rate. For action scenes however, the animators shoot 'on ones' (i.e., 24 fps), and some frames are even shot with multiple exposures, producing a cinematic, faux motion, blur effect.

In fact, OpenGL version 1.1 makes use of an accumulation buffer which is able to simulate effects such as motion blur by accumulating several redraws into a single frame,[3] although this is *not* possible with OpenGL ES. Since this is quite a costly operation in terms of processor speed and memory, it's not commonly used in games. It is illustrated in Figure 3.1.

Figure 3.1 Simulation of a motion blur multiple exposure by accumulating frames in a single buffer

3.3.3 Frame Rate In Practice

As illustrated above, it's perfectly possible to produce convincing movement at 12 fps, but rapid movement requires more frequent updates in order to avoid a juddering effect.

The frame rate required to provide a smooth game experience varies according to game genre. A simple puzzle game such as a two-dimensional chess game may require updates to a small part of the screen at a time, whereas a 3D first person shooter may require the whole screen to be redrawn because the viewing point has changed.

Frames per second in games may also vary based on the complexity of a game (or a particular scene) and by the hardware, especially on platforms where the hardware is not standardized, such as a PC. It's common to have a trade-off between performance and detail.

---

[3] More about motion blur using an accumulation buffer in OpenGL can be found at **fly.cc.fer.hr/~unreal/theredbook/chapter10.html**.

In practice, 24–30 fps is around the minimum frame rate required to represent a smooth immersive world for any game with significant motion, especially 2D scrolling games and 3D games. But in terms of achieving smooth updates, synchronization to the display hardware becomes more relevant than raw frame rate. It helps to understand a little bit more about display technology to complete this discussion about frame rate.

The LCD Display Rate

An important factor connected to frame rate is how often the frame buffer (described in more detail in section 3.4) gets drawn to the screen, which is independent of the frame rate of the game, and is set by a fixed clock on the LCD controller. There is little point optimizing a game to render 120 fps if the LCD only refreshes at 60 Hz (unless using an accumulation scheme as mentioned in section 3.3.2).

As mobile handsets require support for more demanding use cases, mobile display controllers are becoming a lot more complex in their operation. For a display device, 60 Hz should be considered a bare minimum because a lower update frequency is known to cause fatigue and eyestrain in many people. A frequency of 70–75 Hz should be acceptable for most users, while for some 85 Hz is much more comfortable. Anything beyond that is unnecessary for a mobile device, and may simply drain battery power.

On mobile handsets, the refresh rate of the LCD controller is significant to the frame rate of a game because the game can only draw a new frame as fast as the display driver can update the screen.

3.4   About Display Memory

On Symbian OS, the term 'frame buffer' usually refers to a region of system-accessible memory used to represent the pixels of the LCD. Early mobile controllers used to work much like CRT displays with the frame buffer being scanned at a fixed frequency by the display hardware.

Many LCD controllers shipping in devices based on Symbian OS v9.2 or v9.3 come with on-board display memory that is not addressable by the CPU. In these cases, a chunk of system mapped memory is allocated for the frame buffer as usual, but the LCD driver may have to be told when to refresh its internal screen buffer. Typically a scheme will be in place whereby only regions of the screen that have changed will be copied in order to reduce the memory bandwidth.

Figure 3.2 demonstrates how a screen (simplified to 8×8 pixels for illustration purposes) might be updated after a line has been drawn in the frame buffer. The illustration shows that only the rows that differ from those in the LCD's memory are copied over.

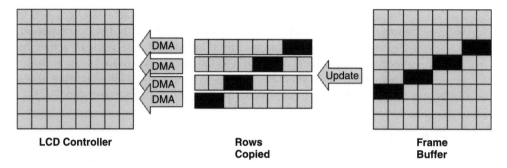

Figure 3.2   How a frame buffer may interact with an LCD display on Symbian OS

Some handsets implement a TV out feature. Since the TV output is generated from the LCD's buffer, what you see on the LCD screen is exactly what appears on the TV (this includes direct drawing to the frame buffer).

A controller may support overlays in hardware and, in some cases, support direct rendering of the camera's viewfinder to the screen in hardware. However, very little of this technology will be exposed directly to third party application or game developers, since it is very handset specific.

The important thing to understand is that as far as most code is concerned, the frame buffer is an area of memory representing a screen which is mapped into all processes allowing any piece of code to modify the contents. The Symbian OS graphics framework takes care of arbitrating all drawing to the screen and is covered in the next section.

3.5   A Primer on Drawing on Symbian OS

Symbian OS provides a UI toolkit based around the concept of a control. In its most simplistic view, a control is a rectangular area of screen represented by a `CCoeControl`-derived class. The control is responsible for issuing the drawing commands required to draw and redraw the area of screen which it owns.

In order for a control to issue drawing commands, it requires a rendering context known as a window. A window is similar in concept to the windows found on desktop systems in that it represents a (potentially overlapping) region owned by an application. When a window needs to be redrawn, the redraw system delegates the drawing request to each control on the window's surface. A control may create its own window and occupy the whole window area or it may use part of its parent window. A control that uses another's window is known as a lodger control, as shown in Figure 3.3.

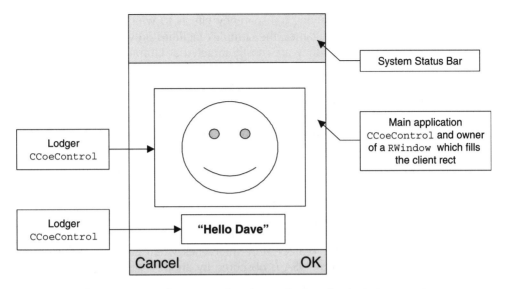

Figure 3.3 A simple representation of an application, showing lodger controls

3.5.1 Reuse of Controls Across UI Platforms

S60, UIQ and MOAP (the UI platform for FOMA phones) all provide their own UI toolkits which contain standard controls for list boxes, menus, dialog boxes and so on, each tailored to the look and feel of their respective platforms. There is little re-usability across platforms; a list box class on S60 is implemented using a completely different control to that used on UIQ.

However, when developing a custom control, the basic CCoeControl framework that Symbian OS provides is common across UI platforms. As long as strong UI dependencies are avoided, then a control can be ported between UI platforms with ease.

Figure 3.3 also serves to illustrate that different parts of the screen can be owned by different processes. The status bar is usually owned by a system process and drawn independently of the application that has the main focus. The system component that sorts out which process needs to draw what, and when, is the Symbian OS window server.

3.5.2 The Symbian OS Window Server

The Symbian OS window server (commonly abbreviated to WSERV) has two key roles. It routes screen pointer events and keypad events to the correct window, and it handles the redrawing of regions of the screen, enabling partial updates of the screen. WSERV has many sophisticated services for rendering UI controls and for establishing transitions between applications, applying transparency effects to windows and so on. But for most kinds of games, the complex facilities provided by WSERV can be ignored since they are mostly targeted at UI frameworks and system code.

The two main types of drawing services that WSERV provides to the games programmer are:

- a graphics context for drawing graphics clipped to a window on screen

- direct screen access (DSA) for custom drawing code.

In the first instance, WSERV takes a very active role in executing and clipping the drawing code to the correct area of screen. For DSA, WSERV takes a back seat, allowing an application to draw what it wants in a specified region. WSERV only intervenes when an area of the screen being managed by DSA is obscured by another window for which it is responsible.

The rest of this chapter deals with writing custom controls. For more in depth information on the workings of WSERV and an introduction to writing GUI applications using the Symbian OS application framework, please see the recent Symbian Press book *Symbian OS C++ for Mobile Phones: Volume 3* by Richard Harrison and Mark Shackman. The book covers in depth how windows, controls, and redraws work on Symbian OS v9. A lot of the information will be relevant if you are writing games which use standard UI controls and elements from UIQ, S60 and MOAP, but it is beyond the scope of this chapter, which is specifically about custom controls. If you are new to Symbian graphics programming, the best way to understand the basic graphics services is to create a new GUI application in the Carbide.c++ IDE. Using the application wizard, find the `Draw()` function, which will have the default implementation of drawing 'Hello World!' to the screen, and start playing around with the different drawing functions available in `CGraphicsContext`.

From a games perspective, it is important to have a basic grasp of how WSERV deals with draws and redraws, since a game has to co-exist with other applications. The next section covers basic drawing.

3.5.3   Drawing a Control

Standard Application Drawing API Overview	
Libraries to link against	`ws32.lib cone.lib`
Header files to include	`ws32.h coecntrl.h`

Standard Application Drawing API Overview

Required platform security capabilities	None
Key classes	CCoeControl, CWindowGc

Each area on the screen where an application wishes to draw must have an associated window. A handle to a window is represented by the class RWindow. It's possible to create a RWindow object from an application or a console-based application, and handle drawing events directly, but for most uses, a CCoeControl-derived custom control is recommended, because it simplifies the API for redrawing to a window.

Typically, a game's main screen will be a CCoeControl-derived class which will either fill the application area (meaning that the status bar is still visible as shown in Figure 3.3) or will fill the entire screen. As soon as the control is made visible, the Draw() function will be called in order to render the initial state of the control. The code fragment below shows how a CCoeControl object clears itself.

```
void CGfxWorkbenchAppView::Draw(const TRect&  aRect ) const
  {
  // Get the standard graphics context
  CWindowGc& gc = SystemGc();

  // Clear the screen
  gc.Clear(aRect);

  }
```

This code introduces a class called CWindowGc, which is a type of graphics context, and will be discussed later in this section. The Draw() function must be quick, and is a const, non-leaving function, which means that it cannot allocate memory or modify the state of the application or game. The binding contract of the Draw() method is that it draws some model or state and, if the state is unchanged between two calls of Draw(), the results should be exactly the same.

A call to Draw() can be initiated by the application itself or by the operating system:

- **Application-initiated drawing**: after the game state has changed, the game can call DrawNow() on the CCoeControl, which results in Draw() being called.

- **System-initiated drawing**: for example, when the stacking order of windows changes so that the game comes to the foreground, Draw() will be called automatically by the system in order to refresh the display.

In fact, on some systems, windows may have their drawing commands cached in WSERV's redraw store, preventing redraw requests from ever getting to the application. For this reason, it is vital that these draw requests don't modify the underlying game state.

3.5.4 Handling a Redraw

Imagine the leftmost screen in Figure 3.4 as a full screen CONE[4] control. Since this is a multitasking device, an incoming event or trigger may cause the control to be obscured at any time as illustrated by the 'Battery Low!' alert dialog shown in the middle screen. Any drawing to the heart control while the battery low indicator is in the foreground will be clipped and so there is no danger of overwriting the pixels of the dialog area. However, when the dialog window closes, shown in the screen on the right, the area previously occupied by the dialog is marked as invalid.

 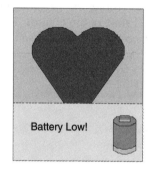 

Figure 3.4   A sequence showing how a region of a control becomes invalid

Invalid regions means that WSERV must locate each window that intersects with that region and make a request to the window's owner to repaint the affected area. In the case shown in Figure 3.4, no other controls are visible and so only the heart control will have its `CCoeControl::Draw(TRect& aRect)` method called.

When a redraw is initiated by the system, the control must ensure that all pixels in the invalid region represented by `aRect` are repainted in order to maintain the integrity of the display. Typically, games render frames to an off-screen bitmap and this makes it trivial to copy only the invalid region back to the screen.

---

[4] The Symbian OS control framework is sometimes referred to as CONE (as a contraction of the phrase 'control environment.') CONE provides a framework for creating user interface controls, a framework for applications to handle user interface events, and environment utilities for control creation and access to windowing functionality.

3.5.5   Graphics Context

In section 3.5.3, we saw basic use of a graphics context in the code. The control simply called `Clear()` in order to blank the control, but there are many more drawing functions provided.

A graphics context is a fairly standard notion across different computing platforms. It encapsulates information about the way things are drawn. A graphics context API is always stateful in that information about things, like pen size and brush color, are built up by issuing separate commands.

The Symbian OS class hierarchy for graphics contexts is shown in Figure 3.5. Class `CGraphicsContext` provides the main interface for drawing geometric shapes and text. Methods are provided for drawing and filling various shapes (points, lines, polylines, arcs, rectangles, polygons, ellipses, rounded rectangles and pie slices), and for drawing text.

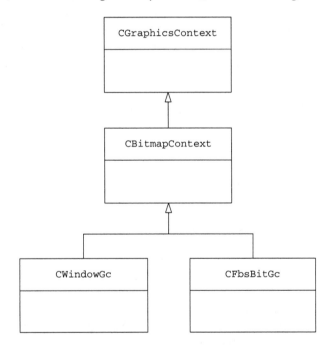

Figure 3.5   Symbian OS graphics context derivation

In practice, all graphics contexts used by a game will be a type of `CBitmapContext`, which is a specialized graphics context adding the following capabilities:

● clearing and copying rectangular areas

● bitmap block transfer.

CWindowGc is the window server client-side API which provides one implementation of a CBitmapContext. Flexible drawing code can be developed by writing against the base class API CBitmapContext. The same API is also used to draw directly to a bitmap based graphics context (CFbsBitGc), so drawing code can easily be re-used from one method to the other.

3.5.6 Window Graphics Context

To understand what the window server does, let's consider the algorithm for drawing a simple line to a bitmap in memory.

When calling CBitmapContext::DrawLine(aStartPoint, aEndPoint), the current pen color and width are retrieved from the graphics context and the pixels corresponding to the line are colored in, overwriting the pixel values of the bitmap as the line is drawn. A bitmap in memory is never obscured (by a window), and so clipping is a matter of containing drawing within the bitmap rectangle.

The algorithm for drawing to a window (via CWindowGc) is more complex:

- a window/control may be partially obscured which requires multiple clipping

- a window may also be semi-transparent and so drawing operations may need to be combined with drawing code for background windows, in order to get the final pixel values for the screen.

Symbian's WSERV implementation contains a drawing command buffer to store commands in the application's process before transferring them in bulk for WSERV to execute. During a Draw(), the command buffer is filled with drawing commands and usually flushed by the UI framework after Draw() has been called on a control. A special optimization has been made for drawing bitmaps. When a bitmap is drawn, only the handle is passed from the application to WSERV. The server looks up the bitmap handle in the shared bitmap server and copies the bitmap in place to the screen.

That's the basics of the Symbian OS window server's drawing mechanism. It's effective for sharing the screen between applications, but in some cases, the mechanism doesn't provide the flexibility and performance required for demanding use cases such as video rendering. To allow for these use cases, Symbian OS supplies a direct screen access API to allow efficient access to the screen, which is covered in the next section.

3.6   Direct Screen Access

DSA API Overview	
Library to link against	`ws32.lib`
Header to include	`ws32.h`
Required platform security capabilities	None
Key classes	`CDirectScreenAccess`, `MDirectScreenAccess`

Game developers wishing to avoid WSERV drawing and composition may use the direct screen access (DSA) API. In a multitasking, highly connected device, direct access to the screen could cause problems with the display content being garbled by multiple applications drawing the same area, effectively scribbling over one another. Symbian's WSERV provides the DSA APIs to marshal access to regions of the screen and manage clients wishing to draw their own content. This section introduces the use cases for DSA, the APIs available and discusses the alternative techniques.

DSA facilitates two modes of access to the screen pixels by providing;

- a `CBitmapContext`-based graphics context that maps directly onto the screen without going through WSERV

- access to the raw screen pixel data (which is useful, for example, for custom bitmap copying routines).

The example code which goes with this section demonstrates how to use both techniques.

The key role of DSA itself is to coordinate clients and ensure that the consistency of the display is maintained. It notifies clients when to stop drawing to a region and when they may resume. This aspect of DSA is very important because the user interface frequently pops up dialogs to announce events such as incoming phone calls or messages (as discussed previously in section 3.5.2).

To use DSA, an application creates a window or control as usual and associates a DSA session with that window. The DSA session provides the information and methods required to draw to the screen. While drawing, a game will usually be exclusively in the foreground. If this situation changes, the DSA session notifies the application of a possible obstruction with an `AbortNow()` callback. When it's safe to resume drawing, a `Restart()` callback is issued and the application can restart the DSA session.

3.6.1 Abort/Restart Sequence

It should be noted that abort or resume callbacks tend to occur in pairs. The sequence of events is typically as follows:

1. The game is in the foreground, drawing graphics in a timer-based loop to the whole screen.

2. The operating system detects an event, such as an incoming call, and requests WSERV to display a dialog.

3. WSERV calls `AbortNow()` on DSA clients.

4. The game synchronously stops drawing.

5. WSERV displays the dialog to indicate an incoming call.

6. WSERV calls `Restart()` on DSA clients.

7. The game updates its drawing code to take into account the new drawing region provided by DSA.

8. Another event is generated. For example, the user ends the call.

9. WSERV calls `AbortNow()` on DSA clients.

10. The game synchronously stops drawing.

11. WSERV removes the incoming call dialog.

12. WSERV calls `Restart()` on DSA clients.

13. The game updates its drawing region and resumes drawing to the whole screen.

As you can see, the process of aborting and resuming the DSA session takes place on each change in window arrangements, and is not just a matter of aborting on appearance and resuming on disappearance.

Well written applications should not continue drawing while they are in the background. Applications may detect being moved to the foreground or background by inheriting from the mixin class `MCoeForegroundObserver` and implementing the functions `HandleGainingForeground()` and `HandleLosingForeground()`. A call to `CCoeEnv::AddForegroundObserverL()` registers the class to receive the callbacks when the application goes into the background and comes into the foreground. The Symbian OS Library in your SDK can give further details on this mechanism.

As Chapter 2 describes, a game should pause when losing focus, and all drawing and timers should halt. This is important to avoid draining the battery.

The following fragments of code show how to construct a DSA session in order to get a graphics context for the screen. The example code forms part of a full project, which is available for download from the Symbian Press website (at ***developer.symbian.com/roidsgame***), the class diagram for which is shown in Figure 3.6.

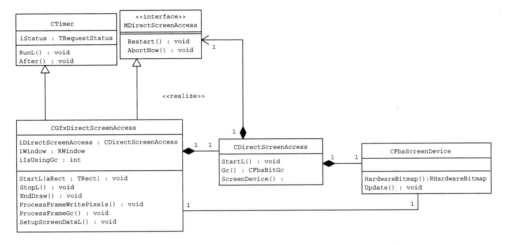

Figure 3.6 A simplified class diagram for the example DSA code

The example contains class `CGfxDirectAccess`, which contains the code below, and also implements a timer to periodically refresh the screen (some of the details of the class are omitted for clarity in the printed version but the code is complete in the version available for download). The DSA session is created with the following code:

```
void CGfxDirectAccess::ConstructL()
  {
  CTimer::ConstructL();
  // Create the DSA object
  iDirectScreenAccess = CDirectScreenAccess::NewL(
      iClient,                // WSERV session
      *(CCoeEnv::Static()->ScreenDevice()), // CWsScreenDevice
      iWindow,                // RWindowBase
      *this                   // MDirectScreenAccess
      );

  CActiveScheduler::Add(this);
  }
```

All of the parameters come from the `CCoeControl` or the UI environment singleton object (`CCoeEnv::Static()`). The final parameter,

`MDirectScreenAccess`, is the class that will receive the `Abort-Now()`/`Restart()` callbacks.

Each time the DSA is required to start drawing to the screen, the following code is called:

```
void CGfxDirectAccess::StartL()
  {
  // Initialize DSA
  iDirectScreenAccess->StartL();

  // This example is hardcoded for a 24/32bpp display
  // See section 3.8 for more information
  if(iDirectScreenAccess->ScreenDevice()->DisplayMode16M()==ENone)
    {
    User::LeaveIfError(KErrNotSupported);
    }

  SetupScreenDataL(); // Obtain the base address of
                      // the screen and other info (see below)

  After(TTimeIntervalMicroSeconds32(KFrameIntervalInMicroSecs));
  }
```

Each time the game loop timer fires, `RunL()` is called, which draws a frame and then re-queues the timer. At this point, it is safe to draw direct to the screen until notified to abort.

```
void CGfxDirectAccess::RunL()
  {
  if(iGcDrawMode)
    {
    ProcessFrameGc();
    }
  else
    {
    ProcessFrameWritePixels();
    }

  EndDraw();
  iFrameCounter++;
  After(TTimeIntervalMicroSeconds32(FrameIntervalInMicroSecs));
  }
```

`ProcessFrame()` is where all the custom drawing code goes. The example code contains two methods for generating content: `Process-FrameGc()`, which uses the direct screen graphics context to draw a few lines and `ProcessFrameWritePixels()`, which writes pixel values directly.

3.6.2 Drawing to the Screen Using a Graphics Context

There are two major advantages of using `Gc()` on a DSA session:

- Standard drawing functions can be used without specializing for screen mode.

- Drawing can be clipped to a region, making it suitable for partially obscured drawing.

The following code demonstrates drawing directly to the screen using the DSA graphics context:

```
void CGfxDirectAccess::ProcessFrameGc()
  {
  CBitmapContext* gc = iDirectScreenAccess->Gc();
  TPoint p1(0,0);
  TPoint p2(0,iDsaRect.Height());
  for(TInt x=0; x<=iDsaRect.Width(); x++)
    {
    TRgb penColor((x+iFrameCounter));
    gc->SetPenColor(penColor);
    gc->DrawLine(p1,p2);
    p1+=TPoint(1,0);
    p2+=TPoint(1,0);
    }
  }
```

Figure 3.7 shows the output to the screen of the previous example code. There is not much difference in this code compared with using a window graphics context.

Figure 3.7 The image resulting from code running on S60 that uses a DSA graphics context. The code draws scrolling green vertical bars

3.6.3 Writing Pixel Data to the Screen

In order to write pixel data directly to the screen, the base address of the screen is required to calculate the address of individual rows and pixels. As with all direct access techniques, manipulating the pixel data requires knowledge of the layout of the data (see section 3.8 for more information).

Over the various releases of Symbian OS, there have been a few ways of retrieving information required to directly address the screen; most of these methods are deprecated or removed. Unfortunately, there is a lot of variability between devices (including the emulator) and manufacturers. The example code below presents a method which should scale across various devices and which also works in the emulator.

The code is encapsulated in the method `SetupScreenDataL()`. It first queries the DSA for a hardware bitmap representing the screen. If the bitmap is not accessible, the screen attributes are retrieved from the HAL (hardware abstraction layer). In practice, the emulator supports hardware bitmaps, but target device support is unlikely. To deal with this, `SetupScreenDataL()` reads and normalizes all values into a data structure, `SScreenData` (defined in the example code), which then allows all the other code to work independently of the implementation.

```
struct SScreenData
  {
  TInt iDisplayMemoryAddress;
  TInt iDisplayOffsetToFirstPixel;
  TInt iDisplayXPixels;
  TInt iDisplayYPixels;
  TInt iDisplayMode;
  TInt iDisplayOffsetBetweenLines;
  TInt iDisplayBitsPerPixel;
  TBool iDisplayIsPixelOrderLandscape;
  };

void CGfxDirectAccess::SetupScreenDataL()
  {
  Mem::FillZ(&iScreenData, sizeof(iScreenData));

  RHardwareBitmap hwBmp =
    iDirectScreenAccess->ScreenDevice()->HardwareBitmap();

  TAcceleratedBitmapInfo bmpInfo;
  hwBmp.GetInfo(bmpInfo);


  // If the platform supports s/w accessible hardware bitmap
  // (emulator?) otherwise get the values directly from the HAL
  if( hwBmp.iHandle && (bmpInfo.iSize.iWidth!=0) )
    {
    iScreenData.iDisplayMemoryAddress = (TInt)bmpInfo.iAddress;
    iScreenData.iDisplayOffsetToFirstPixel = 0;
    iScreenData.iDisplayXPixels = bmpInfo.iSize.iWidth;
    iScreenData.iDisplayYPixels = bmpInfo.iSize.iHeight;
    iScreenData.iDisplayBitsPerPixel = 1 << bmpInfo.iPixelShift;
    iScreenData.iDisplayOffsetBetweenLines = bmpInfo.iLinePitch;
    }
  else
    {
    // The display mode must be set as the incoming paramater for
```

```
// all other values to return correct figures
// The display mode is used to retrieve all the other params

User::LeaveIfError(HAL::Get(HALData::EDisplayMode,
                           iScreenData.iDisplayMode);
// Incoming parameters to HAL must now contain iDisplayMode
iScreenData.iDisplayOffsetToFirstPixel =
iScreenData.iDisplayMemoryAddress =
iScreenData.iDisplayXPixels =
iScreenData.iDisplayBitsPerPixel = iScreenData.iDisplayMode;

User::LeaveIfError(HAL::Get(HALData::EDisplayMemoryAddress,
                           iScreenData.iDisplayMemoryAddress));
// Repetitive code is omitted.
// HAL is used to fill the SScreenData structure...

}
```

The following code extract demonstrates how to fill the screen with RGB values based on screen mode of ECoor16MU. In the code below, iDsaRect defines the screen coordinates of the rectangle representing the window under DSA control and iFrameCounter is simply an integer that gets incremented on each frame. The code simply varies the RGB values of each pixel by using the iFrameCounter variable to influence the color.

```
void CGfxDirectAccess::ProcessFrameWritePixels()
  {
  TUint32* screenAddress =
      (TUint32*)(iScreenData.iDisplayMemoryAddress);

  const TInt limitX = iDsaRect.Width();
  const TInt limitY = iDsaRect.Height();

  TInt rowWidthInBytes = iScreenData.iDisplayOffsetBetweenLines;

  const TInt strideInWords = iScreenData.iDisplayOffsetBetweenLines>>2;
  TUint32* lineAddress = screenAddress;
  lineAddress += (strideInWords*iDsaRect.iTl.iY);

  for(TInt y=0; y<limitY ; y++)
    {
    TUint32* p = lineAddress + iDsaRect.iTl.iX;
    for(TInt x=0; x<limitX; x++)
      {
      TRgb rgb((y+iFrameCounter);
      *(p++)=rgb._Color16MU();
      }
    lineAddress += strideInWords;
    }
  }
```

For each row in the DSA rectangle, the address of the first pixel is calculated and stored in variable p which is used to write 32-bit values

Figure 3.8 A screenshot from the DSA pixel write example. The code draws scrolling red horizontal bars

representing the RGB of each pixel. Variable p is incremented by 32 bits each time so that it always points to the next pixel. This is illustrated in Figure 3.8.

3.6.4   Updating the Screen Device

As I mentioned in section 3.3.3, the LCD may have an internal screen buffer separate from the frame buffer exposed by DSA. When not using the graphics context, the screen device must be notified of which regions of the frame buffer have been modified.

```
void CGfxDirectAccess::EndDraw()
  {
  TRegionFix<1> reg(iDsaRect); // Create a one rectangle region
  iDirectScreenAccess->ScreenDevice()->Update(reg);
  iClient.Flush();
  }
```

Update should always be called to ensure that the Symbian frame buffer is copied to the LCD memory (on systems with such hardware). On the emulator, calling Update() causes the frame buffer to be flushed to the underlying Microsoft Windows GUI window.

3.6.5   Handling a Change in Windows

When the DSA session was created, it required an MDirectScreenAccess derived object to be supplied to the constructor. The MDirectScreenAccess mixin contains two callbacks, AbortNow() and Restart(), as previously described.

Once an `AbortNow()` callback is invoked, the DSA session becomes invalid (including any GC and screen device objects obtained from `CDirectScreenAccess`) and must not be used any more.

```
void CGfxDirectAccess::AbortNow
        (RDirectScreenAccess::TTerminationReasons aReason)
  {
  Cancel();
  }
```

When it is safe to resume drawing, `Restart()` will be called and the client should call `CDirectScreenAccess::StartL()` in order to set up a new DSA and GC. Note that `StartL()` can leave, for example, through lack of memory. If a leave occurs, direct screen access cannot be restarted. For this reason, calling `StartL()` from `Restart()` requires it to be called within a `TRAP()`.

```
void CGfxDirectAccess::Restart
        (RDirectScreenAccess::TTerminationReasons aReason)
  {
  if(iGcDrawMode)
    {
    TRAP_IGNORE(StartL()); // absorb error
    }
  }
```

In the example code, the DSA is only restarted if the GC drawing mode is used since GC drawing supports clipping (the direct mode would overwrite overlapping windows).

`StartL()` recalculates the clipping region so that if direct screen access was aborted because another window appeared in front of it, that window will not be overwritten when direct screen access resumes. Dealing with partial redraws when writing pixels to the screen is very complex and offers little gain. For a game, it's wise to pause and let the user resume. In the example code, the timer is canceled, which stops the redraw.

Note that no other window server functions are called within `Abort-Now()`, because that would cause a dead lock, since the window server session is synchronously waiting. Calls to window server APIs can be resumed when the `Restart()` callback comes through.

3.7 Double Buffering and Anti-Tearing

Double buffering is an established technique for drawing flicker-free graphics. It is an essential idiom used for rendering fluid motion in anything from drag and drop puzzle games to high speed shoot 'em ups. This section discusses the problem that double buffering solves and introduces the concept of tearing, and the methods of preventing tearing.

3.7.1 Buffering Techniques

Table 3.1 shows the buffering techniques available on Symbian OS.

Table 3.1 Buffering schemes and their implementation on Symbian OS

Buffer scheme	Description	Implementation on Symbian OS
Single buffering	A single frame buffer of pixel data is used to draw to the screen.	Direct screen access (DSA), although on systems that require an Update() to flush the frame buffer, the results will be double buffered to an extent.
Double buffering	This scheme uses an off-screen buffer, also known as a back buffer. Double buffering allows slower drawing operations to be performed on the off-screen buffer. The contents of the off-screen buffer can then be quickly copied, or 'blitted', to the current frame buffer, which is also known as the front buffer.	Off-screen bitmap. For synchronized drawing with anti-tearing, it's best to use CDirectScreenBitmap.
Page flipping	Page flipping avoids the need to copy pixel data from one buffer to another. It works by swapping the addresses pointing to the back buffer and the front buffer. Page flipping allows the CPU to start modifying the pixel data in the back buffer as soon as the flip occurs but, once the CPU has finished drawing to the back buffer, it must wait for the vertical blank before it can continue.	Page flipping may be implemented on specific hardware versions of CDirectScreenBitmap.

3.7.2 The Off-screen Bitmap Technique

The most common and portable technique for double buffering is to use an off-screen bitmap (also known as a back buffer) owned by the game. The off-screen bitmap is where the background, tiles, and elements such as high score are combined for each frame and then copied to the screen in one go. The process of copying is often referred to as a 'blit'.

The following code shows how the game **Roids**, shown in Figure 3.9, creates an off-screen bitmap and a graphics context that allows all the usual drawing commands, such as line, fill, and blit, to be executed on the off-screen buffer rather than in the window.

Figure 3.9 **Roids** makes use of an off-screen bitmap

The off-screen bitmap is stored in the variable `iBackupBitmap` of the game's CONE control object, and is of type `CWsBitmap`, a class which specializes `CFbsBitmap` so that the underlying bitmap data can be efficiently shared between the client and WSERV. It's not mandatory to use `CWsBitmap`, but it's faster than `CFbsBitmap` when using a window graphics context.

```
iBackupBitmap = new (ELeave) CWsBitmap(iCoeEnv->WsSession());
error=iBackupBitmap->Create(Rect().Size(),
           iEikonEnv->DefaultDisplayMode());
```

Two more member variables, `iBitmapDevice` and `iBitmapGc` are required in order to establish a graphics context over the bitmap.

```
iBitmapDevice = CFbsBitmapDevice::NewL(iBackupBitmap);
User::LeaveIfError(iBitmapDevice->CreateContext(iBitmapGc));
```

Once successfully created, `iBitmapGc` can now be drawn to instead of `SystemGc()`. When the game model changes, the bitmap can be updated and a `DrawNow()` command issued to the control. The CONE drawing code is very simple since all it has to do is copy any invalidated regions of the `CCoeControl` straight from the bitmap.

```
void CRoidsContainer::Draw(const TRect& aRect) const
  {
  CWindowGc& gc=SystemGc();
  gc.BitBlt(aRect.iTl, iBackupBitmap, aRect);
  }
```

A good tip when creating drawing code is to divide the drawing of individual elements into methods which take a CGraphicsContext rather than using a specific GC. This approach ensures that the graphics context can easily be varied, which is useful for switching between single and double buffering to help debug drawing code. If you have a lot of existing code, it's possible to use CCoeControl::SetGc() to divert calls to SystemGc() to the off-screen bitmap.

The **Roids** code extract below shows how a function can be structured to draw to an arbitrary graphics context.

```
void CRoidsContainer::DrawExplosionParticle(CGraphicsContext& aGc,
                     const TExplosionFragment& aFragment) const
  {
  const MPolygon& shape = aFragment.DrawableShape();

  TInt gray = 255-aFragment.Age();
  aGc.SetPenColor(TRgb(gray, gray, gray));
  aGc.DrawLine(shape.PolyPoint(0), shape.PolyPoint(1) );
  }
```

3.7.3 Tearing

Tearing occurs when the contents of the screen is refreshed part way through drawing the next frame of animation, which is clearly undesirable. Tearing appears on-screen as a horizontal discontinuity between two frames of an animation, as illustrated in Figure 3.10. The application has drawn frame n-1 and the display hardware is part way through

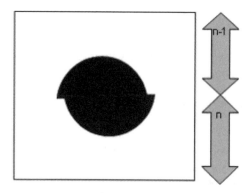

Figure 3.10 Example of two game frames appearing half and half, causing noticeable tearing

refreshing the frame buffer when the application begins to render frame n. A synchronization step is missing which results in a page tearing effect.

If significant, the tearing can be very distracting to a player and ultimately can contribute to eye strain or fatigue. To avoid tearing and provide smooth updates, the display must be synchronized. For smooth rendering of graphics without any tearing artefacts, all changes to the frame buffer should be done during the short period after the display has been refreshed and before the next refresh.

Anti-tearing API Overview	
Library to link against	`scdv.lib`
Header to include	`cdsb.h`
Required platform security capabilities	None
Key classes	`CDirectScreenBitmap`, `TAcceleratedBitmap Info`

Symbian OS provides an API called `CDirectScreenBitmap`, which provides double buffering which (depending on the phone hardware) may be synchronized to the LCD controller. It's often called the anti-tearing API since it's the only documented way to synchronize screen updates with the LCD display. It's most commonly used for video rendering and high speed graphics.[5] The **Roids** example uses the API as shown in Figure 3.11. The mechanism works like this:

1. The game opens a DSA session with the window server as described in section 3.6.

2. The game asks for a `CDirectScreenBitmap` object for a portion of the screen (a rectangle).

3. The game tells the API when the bitmap has finished being drawn to and asynchronously requests for it to be displayed.

4. The game gets called back on the next 'vsync' and can start modifying the bitmap data again.

5. Loop to 3.

---

[5] `CDirectScreenBitmap` relies on a specific implementation to be present on the target hardware. For a given device, there is no guarantee that such an implementation is available.

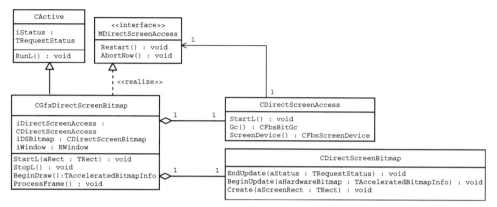

Figure 3.11 Simplified representation of the direct screen bitmap example class

One thing to note is that, in using `CDirectScreenBitmap`, the LCD controller takes control of refreshing the display at the right moment. This may vary from device to device and will be dependent on the LCD display refresh rate, so it's not wise to use a game tick timer. It's also worth mentioning that the callback is not guaranteed to occur on the next vertical sync, since other higher priority threads may pre-empt the application at any time.

Initializing the bitmap is very simple when used with a DSA session. All that's required is the rectangle representing the portion of screen which will be drawn directly.

```
void CGfxDirectScreenBitmap::StartL(TRect& aRect)
  {
  iControlScreenRect = aRect;
  User::LeaveIfError(iDSBitmap->Create(iControlScreenRect,
                     CDirectScreenBitmap::EDoubleBuffer));

  //Initialize DSA
  iDirectScreenAccess -> StartL();

  ProcessFrame();
  }
```

Each time the bitmap needs to be updated, the `BeginUpdate()` function must be called, which returns the bitmap information and memory address of the bitmap area.

```
TAcceleratedBitmapInfo CGfxDirectScreenBitmap::BeginDraw()
  {
  TAcceleratedBitmapInfo bitmapInfo;
  iDSBitmap->BeginUpdate(bitmapInfo);
  return bitmapInfo;
  }
```

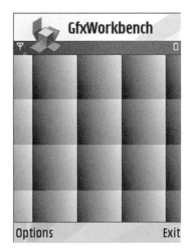

Figure 3.12 Manipulation of the display pixels using direct bitmap

In this simple example, ProcessFrame() gets the address of the bitmap and writes some RGB values directly into the bitmap data – the resulting output is shown in Figure 3.12. The code is similar to the DSA example described in section 3.6, but, this time, the bitmap coordinates start from coordinate (0,0). Note that the following code assumes a 24 bits per pixel (bpp) bitmap with each pixel stored in 32 bits.

```
void CGfxDirectScreenBitmap::ProcessFrame()
  {
  TAcceleratedBitmapInfo bmpInfo = BeginDraw();

  TUint32* lineAddress = (TUint32*)bmpInfo.iAddress;

  TInt width = bmpInfo.iSize.iWidth;
  TInt height = bmpInfo.iSize.iHeight;

  // Code assumes 4 byte words
  const TInt rowWidthInWords = bmpInfo.iLinePitch>>4;

  TInt y=height;
  while(--y)
    {
    TUint32* p=lineAddress;
    for(TInt x=0; x<width; x++)
      {
      TRgb rgb(255,
        (x+iFrameCounter)%64 <<2,
        (y+(iFrameCounter/2))%64 << 2, 0);

      *(p++)=rgb._Color16MU();

      }
    lineAddress += rowWidthInWords; //  go to next row
    }
```

```
iFrameCounter++;
EndDraw();
}
```

All drawing is done to the back buffer and nothing is reflected on screen until the call to: `EndDraw()`.

```
void CGfxDirectScreenBitmap::EndDraw()
{
if(IsActive())
  {
  Cancel();
  }
iDSBitmap->EndUpdate(iStatus);
SetActive();
}
```

`CDirectScreenBitmap::EndUpdate(iStatus)` starts the operation to update the screen. This asynchronous method completes when the display has been updated, to indicate that the bitmap can be reused to draw the next frame.

3.7.4 A Waste of Time?

One issue with waiting for synchronization is that the waiting time is potentially wasted. The game cannot do any rendering to the direct bitmap until its thread's active scheduler has processed the request and called `RunL()`. It is not uncommon to ignore the vsync callback and tune the drawing code for the hardware. Another solution is to use triple buffering, so that there is another back buffer to render to, while the screen refreshes. Triple buffering is usually a bit excessive in terms of memory, and it requires an extra blit.

3.8 Pixel Color Representations

In computer graphics, a bitmap is most commonly represented by a two-dimensional array of picture elements (pixels) which can be presented to an output device for display to a (human) viewer. Since bitmaps are usually closely tied to the display hardware, there are a plethora of different formats. Mobile phones, like desktop PCs, have supported a various screen sizes and color depths over the years. For example, the Psion Series 5 and Series 5MX devices (which were based on an antecedent of Symbian OS as it is today) used 4 and 16 shades of gray respectively, whereas the latest smartphones support 18-bit or 24-bit color displays.

Each pixel in a bitmap is a representation of the color that will ultimately be displayed on a screen pixel. This can either be an index to a palette entry (common with 256 or fewer color displays) or a vector in a color space. The most common color spaces are grayscale intensity, RGB, or YUV. YUV is derived from television standards and is often used as the output from video codecs and camera hardware, but is seldom used in games. Symbian OS graphics interfaces support indexed palettes, grayscale, and various RGB color modes, with RGB being, by far, the most common format used in games.

Table 3.2 shows the different formats allowed for both screens and bitmaps.

Table 3.2 Screen and bitmap display modes available on Symbian OS

Display Mode	Bits per pixel (bpp)	Notes
EGray2	Monochrome display mode (1 bpp).	Chiefly used for binary masks (i.e., one bit to denote if a pixel is opaque or transparent).
EGray4	Four grayscales display mode (2 bpp).	Not commonly used.
EGray16	16 grayscales display mode (4 bpp).	Not commonly used.
EGray256	256 grayscales display mode (8 bpp).	Used primarily for alpha channel masks, where 0–255 grayscale represents an increasing opacity.
EColor16	Low color EGA, 16 color display mode (4 bpp).	Not commonly used.
EColor256	256 color display mode (8 bpp).	Websafe or Netscape palette, and intermediate gray levels. Chiefly used with a palette when loading icons and animated GIFs.
EColor64K	'64,000' 65536 color display mode (16 bpp: RGB565).	Native mode for 16 bpp display.

(*continued overleaf*)

Table 3.2 (*continued*)

Display Mode	Bits per pixel (bpp)	Notes
EColor16M	True color display mode (24 bpp: RGB888).	Not used in practice. It is more compact than 16MU (3 bytes per pixel, rather than 4 bytes) but less efficient to process as the number of bits per pixel is not a power of 2. Displays rarely support its use.
EColor4K	4096 color display (12 bpp: XRGB4444).	Native mode for 12 bpp displays (such as the Nokia 9210).
EColor16MU	True color display mode (32 bpp: XRGB8888).	Common display format for 18 bpp and 24 bpp displays (but the top byte is unused and unspecified). 18 bpp hardware takes the 24 bpp data which software produces and dithers on the fly/drop the two least significant bits of each color component.
EColor16MA	Display mode with alpha (32 bpp: ARGB8888 – 24 bpp color plus 8 bpp alpha).	Only used for bitmaps with alpha channels which need to be combined with transparency on screen.

Note that the display mode of a bitmap in memory is independent of the screen. Graphics context functions such as BitBlit() will automatically convert a bitmap to the depth of the target – be it another bitmap or the frame buffer.

3.8.1 Bitmap Drawing (Blitting) Performance

The availability of different screen modes shown in Table 3.2 means that different depth bitmaps can be combined as they are drawn. There is nothing stopping a game from storing tiles in 256 colors, blitting them to a 4K off-screen bitmap and then copying the bitmap to a 24 bpp display. Having said that, there is an implicit performance penalty when blitting from a source to a destination with a different depth. The penalty may be significant, for example, when porting code which is optimized for a particular screen mode. In order to quantify the cost, a small profiling app was written which:

- creates a bitmap in each of the formats listed in the table above

- blits the bitmap several times and takes the mean time for each display format.

The results, shown in Figures 3.13 and 3.14, show the frames per second achieved for each format on two different Symbian smartphones, the Nokia E61 and Nokia N95.[6]

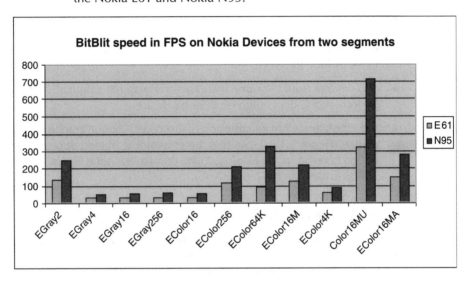

Figure 3.13 Blit speed in fps for each bitmap format on hardware

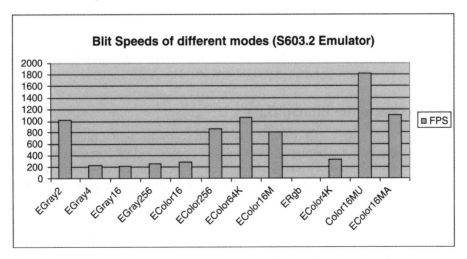

Figure 3.14 Blit speed in fps for each bitmap format on the S60 3.2 emulator

---

[6] I've discussed the results in an article called *Blit speed and bitmap depth,* which can be found on my blog at ***twmdesign.co.uk/theblog/?p=52***.

The result that stands out from the data collected is that the default display format for both the devices and the emulator is EColor16MU; the graphs clearly show a drastic performance increase when using this mode. This leads to the following conclusions:

- the bitmap depth should match the screen when performance is a consideration

- there is a trade off between memory and speed. Storing game tiles as 256 bpp index modes may use only one quarter of the memory compared to the equivalent EColor16MU, but it takes almost three times as long to render

- two devices from the same manufacturer may have radically different performances, depending on their hardware

- the emulator is not reliable for realistic data about graphics performance. Figure 3.14 shows that the emulator is blitting approximately 600 % faster in mode EColor16MU compared to the Nokia E61 (and 260 % faster than the Nokia N95).

3.8.2 Grayscale Images as Masks

For grayscale images, the pixel value is a single number that represents the pixel's brightness. The most common pixel format is the byte image, where this number is stored as an 8-bit integer giving a range of possible values from 0 to 255. Typically, zero is taken to be black, and 255 taken to be white; values in between make up the different shades of gray.

Grayscale images (mode EGray256) are most commonly used as alpha masks for specifying transparency when using functions such as CBitmapContext::BitBltMasked().

3.8.3 Access to the Bitmap Data

A CFbsBitmap is a client-side handle to the actual bitmap residing in the font and bitmap server's heap. The font and bitmap server (known as FBServ for short) has two heaps: a small heap and a large heap. Bitmaps of less than 16 KB (previously 4 KB in Symbian OS v9.2 and earlier releases) are stored in the small heap, and bitmaps larger than this threshold are stored on the large heap.

The reason for having two heaps is to make it easier for the system to reclaim space made available by an application freeing a large bitmap object, by minimizing heap fragmentation. Fragmentation occurs when large bitmaps are de-allocated, leaving holes in the heap. Symbian

OS currently only supports shrinking heaps from the end and, without defragmentation, the memory ocupied by those holes effectively becomes unusable.

To avoid potential waste, at certain safe points, FBServ will reclaim the memory by shuffling up the bitmaps on the large heap in order to fill any holes. The memory committed to the heap can be reduced by shrinking from the end. The heap should be locked while pixel data is being read or manipulated, since the actual memory location of the pixel data may otherwise change as the heap is defragmented.

There are three ways of accessing the pixel data:

- `TBitmapUtil` – safely reads and writes a pixel at a time

- `CFbsBitmap::GetScanLine()/SetScanLine()` – copies a scan line into a buffer that can be accessed and modified before being put back in the original

- `CFbsBitmap::DataAddress()` – returns the memory location of the pixel data for direct manipulation of the bitmap in place.

The first two methods are safe, since they lock the large heap if necessary, but they may not be as efficient for all bitmap manipulation tasks. Let's consider each of the methods in more detail.

TBitmapUtil

Roids uses `TBitmapUtil` to draw single pixels onto the off-screen bitmap, as illustrated by Figure 3.15. During development, I discovered that use of `TBitmapUtil` is negligible on the frame rate for plotting a

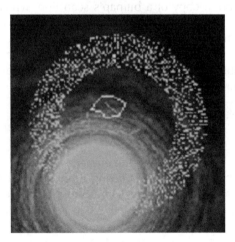

Figure 3.15 Starburst from the game ***Roids*** are drawn using `TBitmapUtil`

thousand or fewer pixels. But I found it to be too slow when processing entire frames or bitmaps in real time.

```
void CRoidsContainer::DrawBackgroundStars(CWsBitmap& aBitmap) const
  {
  TBitmapUtil bUtil(&aBitmap);

  TDisplayMode display = iEikonEnv->DefaultDisplayMode();

  bUtil.Begin(TPoint(0,0));

  for(TInt i=0; i<EMaxStars; i++)
    {
    TPoint point(iStars[i].iX, iStars[i].iY);
    bUtil.SetPos(point);

    TInt gray = i%200 + 55; // A simple calculation to vary the color
    TRgb rgb(gray,gray,gray);
    bUtil.SetPixel(rgb._Color16MU());
    }

  bUtil.End();
  }
```

Note that code must explicitly deal with different color depths when setting a pixel value. The call to _Color16MU works in this instance, but will produce garbled colors on 16 bpp displays.

CFbsBitmap::GetScanLine() and *CFbsBitmap::SetScanLine()*

Manipulating scan lines is simple, safe, and fairly efficient (it's essentially just a Mem::Copy() when the display modes match, which is an optimized memory copy). The approach uses an intermediate buffer to store a copy of a bitmap's scan line, which can be manipulated at the application's leisure (no need to keep a lock), and then returned to the bitmap.

In order to modify a scan line, an application must first create an 8-bit buffer that is large enough to take a full scan line (this will depend on the screen mode).

```
void GetScanLine(TDes8& aBuf, const TPoint& aPixel, TInt aLength,
                          TDisplayMode aDispMode) const;
```

The method gets the bitmap's scan line for a line specified by a point, aPixel, and by retrieving aLength pixels into the buffer aBuf, then converting it (if necessary) to display mode aDispMode. Once the scan line has been processed, it can be returned with a call to the following:

```
void SetScanLine(TDes8& aBuf, TInt aY) const;
```

CFbsBitmap::DataAddress()

`CFbsBitmap::DataAddress()` can be used to read and write pixel data in a similar way to the DSA examples shown in section 3.6. As with DSA, the layout of the data must either be assumed or can be inferred from the display mode. Direct manipulation of a bitmap's pixel data should be protected, by locking the heap, using the following boilerplate code:

```
CFbsBitmap* myBitmap;
myBitmap->LockHeapLC() // Push a lock onto the cleanup stack
TUint32* dataAddress = myBitmap->DataAddress();

// manipulate bitmap data calling leaving code
// ...

CleanupStack::PopAndDestroy(); // Unlock heap
```

If the manipulation code doesn't leave, then `LockHeap()` and `UnlockHeap()` may be used instead of the leaving versions shown above.

Whenever the bitmap is read from or written to, you should ensure that:

- the code between lock and unlock is short, since other parts of the system will be waiting on the release of the heap lock in order to draw bitmaps

- no nested calls are made to `LockHeap()`. This can happen when other `TBitmapUtil`, `CFbsBitmap`, or icon drawing methods are called within the same thread

- when using `LockHeap()`, if a leave can occur, `UnlockHeap()` must be called where it is handled, since if it is not, the heap will stay locked. This can be effected using the cleanup stack.

Note that system bitmaps (usually located in the ROM) may be compressed, which means that the data will be run-length encoded (RLE). Knowing this may save you time, if you are using a system bitmap for convenience in prototype code. Application and game bitmaps are uncompressed in RAM unless `Compress()` is explicitly called on the bitmap.

3.9 Loading and Manipulating Images

Images (often called bitmaps) are hugely important to games. They provide the frames of the animations, and the fonts, backdrops, and textures that make up the look and design of the game. Images are often created by a dedicated graphics artist and are usually tuned to the screen resolution and depth of the target platform. Symbian smartphones support many of the popular image formats; Table 3.3 compares the compression mode and utility of the common formats.

Table 3.3 Compression ratios for popular web graphics formats

Format	Typical compression ratios	Description
MBM	Very dependent on bitmap data.	MBM supports RLE encoding for up to 32 bpp bitmaps, which works well for simple bitmaps and masks, where runs of a similar color are common. The MBM is only compressed if a ratio of 1.25:1 can be achieved. MBM is a very convenient format for icons, color font, game tiles, sprites and animations (with masks). MBM supports multiple images per file which makes the format suitable for simple animation.
GIF	4:1–10:1	Lossless for images ≤ 256 colors. Works best for flat color, sharp-edged comic or icon art, but more complex images tend to require dithering. Animated GIFs are very useful for cartoon sequences in games.
JPEG (High)	10:1–20:1	High quality, with little or no loss in image quality with continuous tone originals. In fact, there is usually little need for such high quality on small screens.

Table 3.3 (*continued*)

Format	Typical compression ratios	Description
JPEG (Medium)	30:1–50:1	This range is ideal for game screens, backdrops and photos.
JPEG (Low)	60:1–100:1	Suitable for thumbnails and previews. Visible blocking artefacts.
PNG	10–30 % smaller than GIFs	PNGs behave similarly to GIFs, with up to 32-bit color and a transparency channel. They work best with flat-color, sharp-edged art. PNGs compress both horizontally and vertically, so solid blocks of color generally compress best. PNG is best suited for titles, large game graphics, detailed animation frames, and typefaces stored as bitmap tiles.

3.9.1 Loading Native Bitmaps

MBMs are the most convenient graphics format on Symbian OS since they support multiple images per file and can be generated as part of the build. Convenient functions exist in the singleton `CEikonEnv` (part of the application environment) to load MBM images synchronously into `CFbsBitmap` objects. In order to generate an MBM, ready for use in a game, a declaration such as the following must be added to the game's MMP file.

```
START BITMAP nebula.mbm
  HEADER
  TARGETPATH    \Resource\Apps
  SOURCEPATH images
  SOURCE c24 nebula.bmp
END
```

The order of the list of `SOURCE` bitmaps is significant and is preserved in the MBM. When the project is built, it will generate two files:

- `epoc32\resource\apps\nebula.mbm` (the binary)

- `epoc32\include\nebula.mbg` (a C++ header file used to reference the contents of the MBM in source and resource files).

Loading the bitmap is as simple as calling with the correct enumeration value as it is defined in `nebula.mbg`.

```
_LIT(KNebulaMbmFilename, "\\resource\\apps\\nebula.mbm");
CWsBitmap* neb =  iEikonEnv->CreateBitmapL(KNebulaMbmFilename(),
                                              EMbmNebulaNebula);
```

The `CWsBitmap` object can then be used to draw to the `SystemGc()` graphics context.

Although this is very useful for development, the technique only works for MBM files which tend to produce large files. To load and display JPG, PNG or GIF files, they must first be decoded using the image conversion library (ICL), which is the subject of the next section.

3.9.2   Image Conversion Library (ICL)

ICL API Overview	
Library to link against	`imageconversion.lib`
Header to include	`imageConversion.h`
Required platform security capabilities	None
Key classes	`CImageDecoder,` `CBufferedImageDecoder`

Symbian supports the encoding and decoding of images using the ICL. The library is designed for converting graphics formats into `CFbsBitmaps` and vice versa.

Multithreaded Support

The ICL supports decoding in a separate thread in order to reduce latency of active objects in the client's thread. This may or may not be important to a game. An example of where background thread decoding would be beneficial is in the case of a game using the Google Maps API, where the map can be dragged around and images updated as they are received over the network. If the codec processes chunks of the image in an active object then it may cause unpredictable delays in servicing other active objects such as pointer or redraw events. This can cause an erratic, 'jittery' user experience. Moving the decoding into a dedicated thread would allow it to continue in the background without impacting the latency of UI events.

The ICL is a multimedia framework which depends on plug-ins assembled by the handset manufacturer. Some of the plug-ins may use hardware acceleration, but that's less common, since hardware may limit the number of simultaneously decoding JPEGs (which happens when web browsing).

Loading Images

The ICL encodes and decodes images asynchronously, so clients must handle a completion event when an image has been decoded. The example code shows how to wrap up an image decoding operation in an active object class called CGfxImageDecoder.

```
class CGfxImageDecoder : public CActive
  {
public:
  CGfxImageDecoder(CFbsBitmap* aBitmap, CFbsBitmap* aMask,
                   MGfxImageDecoderHandler& aHandler);
  ~CGfxImageDecoder();
  void LoadImageL(const TDesC& aFilename);
protected:
  void RunL();
  void DoCancel();
private:
  CImageDecoder* iDecoder;
  CFbsBitmap* iBitmap; // not owned
  MGfxImageDecoderHandler& iHandler;
  RFs iFs;
  };
```

The decoder is constructed by passing a pointer to an empty bitmap handle (one which has not had Create() called on it yet). The example code uses a MGfxImageDecoderHandler mixin to signify image-loading completion via a callback.

```
CGfxImageDecoder::CGfxImageDecoder(CFbsBitmap* aBitmap, CFbsBitmap*
                                 aMask, MGfxImageDecoderHandler& aHandler)
: CActive(0),
  iBitmap(aBitmap),
  iHandler(aHandler)
  {
  CActiveScheduler::Add(this);
  }
```

In order to load an image, the decoder is instantiated with the image file using CImageDecoder::FileNewL() and queried in order to get the size and depth of the image as a TFrameInfo object.

```
void CGfxImageDecoder::LoadImageL(const TDesC& aFilename)
  {
  User::LeaveIfError(iFs.Connect());

  iDecoder = CImageDecoder::FileNewL(iFs, aFilename);
  TFrameInfo info = iDecoder->FrameInfo();

  // Match the format with an in memory bitmap
  User::LeaveIfError( iBitmap->Create(info.iOverallSizeInPixels,
                                      info.iFrameDisplayMode) );

  iDecoder->Convert(&iStatus, *iBitmap);
  SetActive();
  }
```

The `TFrameInfo` values are used to create a new bitmap to match the size and depth of the image. It is possible to supply different sizes and depths, but this will cause the conversion process to reduce the number of colors and/or scale down the image to fit the supplied bitmap. If the image needs to be converted to the current display mode anyway, then it's best to supply for the current screen mode when creating the bitmap. The final step is to initiate the `Convert()` and wait for the operation to complete asynchronously. On completion (successful or otherwise), `RunL()` is called and, in this example, the status is passed directly onto the handler.

```
void CGfxImageDecoder::RunL()
  {
  iHandler.GfxImageLoadedCallBack(iStatus.Int());
  }
```

Using the Graphic Loading Wrapper

The `CGfxImageDecoder` described above is trivial to use. A view or `CoeControl`-derived object can use the following code to start loading the image (assuming the image file `robot.jpg` exists on the C: drive).

```
void CGfxIclView::StartL()
  {
  iLoadingBmp = new(ELeave) CFbsBitmap();
  iImageDecoder = new(ELeave) CGfxImageDecoder(iLoadingBmp, NULL,
                                               *this);
  _LIT(KRobotFileName, "c:\\data\\images\\robot.jpg");
  iImageDecoder->LoadImageL(KRobotFileName);
  }
```

The following method is called on completion:

```
void CGfxIclView::GfxImageLoadedCallBack(TInt aError)
  {
```

```
iBitmap = iLoadingBmp;
iLoadingBmp = NULL;
DrawNow();
}
```

The `Draw()` function only attempts to draw the bitmap when it has successfully loaded (hence the use of `iLoadBmp` as a temporary handle). This way `Draw()` can be safely called while the JPG is loading – the control will be cleared as seen in the code below.

```
void CGfxIclView::Draw( const TRect& /*aRect*/ ) const
{
// Get the standard graphics context
CWindowGc& gc = SystemGc();

// Get the control's extent
TRect drawRect( Rect());

// Clear the screen
gc.Clear( drawRect );

if(iBitmap)
  {
  gc.BitBlt(TPoint(0,0), iBitmap);
  }
}
```

Figure 3.16 shows a JPG loaded using the sample code above.

Figure 3.16   A JPG image of a robot loaded using the ICL as shown in the sample code

Pre-loading Multiple Images

Games tend to contain a lot of graphics, and processing all the graphics files is usually a large contributor to the time taken to load a game (or a level in a game). The example above can easily be extended to pre-load a list of bitmaps and call the handler for each image (in order to update a progress bar) and then make a final callback when the whole batch operation is complete.

Once the ICL has delivered a CFbsBitmap, it can be used like any other MBM: copied to the screen, scaled, rotated, used as a mask, or even supplied as a texture to OpenGL ES.

3.9.3 Techniques for Using Tiles and Animation Frames

Tiles and animation frames can be implemented using multi-image formats such as MBM or GIF. But it's also possible to store multiple tiles in a single image and copy a section of decoded image using the BitBlt() overload which takes a source rectangle. Figure 3.17 shows how the game tiles can be stored in a flat bitmap and assembled into a game screen, and the code fragment below illustrates how the bitmap in Figure 3.17 could be used to copy part of a bitmap to the graphics context. It blits tile tileNumberToDraw to the destination graphics context.

```
// set a rectangle for the top-left quadrant of the source bitmap
const TInt KTileWidthInPixels = 16;
const TInt KTileHeightInPixels = 16;
const TInt KNumTilesPerRow = 4;

TInt x = (tileNumberToDraw % KNumTilesPerRow) * KTileWidthInPixels;
TInt y = (tileNumberToDraw / KNumTilesPerRow) * KTileHeightInPixels;

TRect sourcRect(TPoint(x,y),
                TSize(KTileWidthInPixels, KTileHeightInPixels));

TPoint pos(100,100); // destination
gc.BitBlt(pos, bitmap, sourcRect);
```

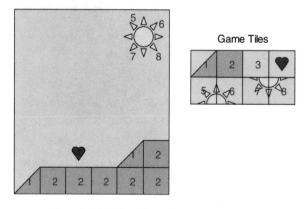

Figure 3.17 Using tiles to draw a game screen

3.10   Scaling for Variable Screen Sizes and Resolutions

Developing a scalable application is a difficult task since it requires some thought about layout geometries and performance variances of a set of devices. Another problem is that a different screen size adds another permutation to test. For certain games, a different screen size can completely change the dynamics, for instance, if a sideways scrolling shooter designed for a QVGA screen attempts to be dynamic and make use of the extra width of an E90 communicator, then the amount of scenery rendered would increase, allowing the user to see much further ahead in the game, perhaps causing a frame rate drop in the game.

There are many ways of designing games which are scalable, for example:

- plan the layout so it can make use of the extra space

- use technologies such as FlashLite and OpenGL ES, which are very attractive when targeting multiple devices, since not only will the 3D rendering be optimized in hardware, but OpenGL ES by its nature can be scaled to any size[7]

- render images to a fixed sized bitmap and scale the bitmap to fill the screen.

Both UIQ 3 and S60 3rd Edition have introduced methods for scaling graphics and making the handling of changes to the screen orientation easier.

3.10.1   Standard Screen Resolutions

It's worth noting that a large pixel resolution may not correspond to a larger physical screen size, it may simply represent a screen with finer pixels. Or, conversely, when comparing two handsets, screens that have different physical screen dimensions may still have the same pixel resolution.

S60 3rd Edition and UIQ 3 have both defined a standard set of screen resolutions supported (and tested) by the respective UI platform (though some specific devices such as the letterbox screen E90 deviate from the platform).

S60 3rd Edition supports two orientations – landscape and portrait – and a number of resolutions: 176×208 (legacy resolution, used in S60 2nd Edition), 240×320 (QVGA), and 352×416 (double resolution). In landscape mode, the soft keys might be located on any side of the display.

---

[7] However, OpenGL ES textures may start to degrade in quality if scaled too large.

UIQ has a more complex matrix of screen configurations based on the presence of absence of touch support, screen orientation and whether the flip (if supported) is open or closed. These are shown in Table 3.4.

Table 3.4 UIQ 3 styles and specifications

Configuration	Screen resolution	Interaction style	Touch screen	Orientation
Softkey style	240 × 320	Soft keys	No	Portrait
Softkey style touch	240 × 320	Soft keys	Yes	Portrait
Pen style landscape	320 × 240	Menu bar	Yes	Landscape
Pen style	240 × 320	Menu bar	Yes	Portrait
Softkey style small	240 × 256	Soft keys	No	Portrait, small

3.10.2 SVG-T

Bitmap formats such as MBMs are suitable for photographic images and simple icons, but tend to look jaggy when scaled either up or down. High contrast edges, in particular, deteriorate when scaled down. Vector-based graphics such as Flash and SVG can be scaled and rotated without loss of quality. The geometric shapes and fills which make up the graphics are recalculated and then rasterized as needed. Vector-based rendering also deals effectively with the problem of screen rotation on a non-square pixel display.

The benefits of SVG over MBMs can be seen clearly in Figure 3.18, where the single shape definition has been drawn with anti-aliasing at several sizes without deteriorating quality.

S60 has supported SVG icons since S60 2.8 and UIQ introduced SVG icon for applications in UIQ 3.1. In S60, SVG icons are created by using the Carbide.c++ IDE, which produces MIF files (an S60-specific way of wrapping up SVG icons). Further information can be found in the Carbide.c++ help files.[8]

Within an S60 application, SVG icons can be loaded using the Akn-IconUtils utility which returns a CFbsBitmap of the rendered SVG, which can be used with the usual graphics context functions. Calling

---

[8] More information about the Carbide.c++ IDE can be found at **www.forum.nokia.com/carbide**.

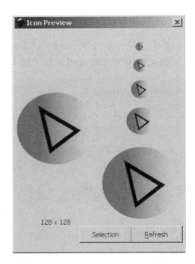

Figure 3.18   The *Roids* application icon as previewed in the SVG editor Inkscape

`AknIconUtils::SetSize()` on the bitmap causes SVG icons to be rendered at the new size.[9]

3.10.3   Dealing with Screen Orientation Changes

Screen orientation is usually discussed and specified in terms of landscape and portrait. When a system orientation change occurs, a lot of processing happens. All applications are notified of the new orientation, and each application may try to recalculate its view and scale icons and controls. After a layout switch occurs, the cursor keys may behave differently, for example, a switch to landscape would result in the up cursor key now generating a right cursor key event.

3.10.4   Explicitly Setting the Orientation

A game will usually be designed explicitly for landscape or portrait. Some handsets may have hardware designs optimized for playing games in landscape mode. Luckily a game doesn't have to adapt to the current orientation, it can explicitly set its mode – forcing an orientation change each time the game is brought to the foreground. The code required to set the orientation explicitly on S60 3rd Edition and UIQ 3 is as follows.

---

[9] More information about using SVG-T on S60 can be found in the document titled *S60 Platform: Scalable Screen-Drawing How-To* which is best accessed by searching on the main Forum Nokia website at *www.forum.nokia.com*. If you're confident of your typing, you can navigate directly to the document at *www.forum.nokia.com/info/sw.nokia.com/id/ 8bb62d7d-fc95-4ceb-8796-a1fb0452d8dd/S60_Platform_Scalable_Screen-Drawing_ How-To_v1_0_en.pdf.html*.

Setting the Orientation in S60 3rd Edition

When `AppUi()->SetOrientationL()` is called explicitly, it over-rides any orientation presently in use by the rest of the phone.

```
void CMyS60App::RotateMe()
  {
  // Changing from portrait to landscape or vice versa.
  iIsPortrait = !iIsPortrait;

  // Change the screen orientation.
  if (iIsPortrait)
    {
    AppUi()->SetOrientationL(CAknAppUi::EAppUiOrientationPortrait);
    }
  else
    {
    AppUi()->SetOrientationL(CAknAppUi::EAppUiOrientationLandscape);
    }
  }
```

Setting the Orientation in UIQ 3

UIQ 3 works differently to S60 3rd Edition when dealing with orientation. UIQ is much more view oriented, and an application can define a separate view resource for each screen mode listed in section 3.10.1.

The following code shows how the orientation can be explicitly set by the application. This in turn selects the most appropriate view defined in the application resource file. Further information is available in the UIQ SDK documentation and from the UIQ Developer Community.[10]

```
void CMyUiqApp::SwapPortLand()
  {
  TQikUiConfig config = CQUiConfigClient::Static().CurrentConfig();
  TQikViewMode viewMode;

  if(config == KQikPenStyleTouchLandscape)
    {
    if( iWantFullScreen )
      viewMode.SetFullscreen();
    else
      viewMode.SetNormal();

    CQUiConfigClient::Static().SetCurrentConfigL
                (KQikSoftkeyStyleTouchPortrait);
    }
  else if(config == KQikSoftkeyStyleTouchPortrait)
    {
```

---

[10] There is a good discussion about how to detect changes of UI configuration at **developer.uiq.com/forum/kbclick.jspa?categoryID=16&externalID=101&searchID=143583** (or search for 'How can I detect changes of UI configuration?') from the main site (**developer.uiq.com**).

```
        viewMode.SetFullscreen();
        CQUiConfigClient::Static().SetCurrentConfigL
                        (KQikPenStyleTouchLandscape);
        }
    iBaseView->SetMyViewMode(viewMode);
    }
}
```

3.10.5   Detecting a Layout Change

Certain types of game may need to adjust to screen size and portrait/landscape changes. The change comes as a notification message which individual controls or views can handle.

Detecting a Layout Change on S60 3rd Edition

Applications can respond to changes in screen size and layout by overriding the `CCoeControl::HandleResourceChange()` method and detecting the `KEikDynamicLayoutVariantSwitch` message.

The code below shows how a full screen control with a DSA drawing engine may deal with a screen rotation. The extent of the control is reset to fill the screen in the new orientation.

```
void CMyScalableDrawingControl::HandleResourceChange(TInt aType)
  {
  CCoeControl::HandleResourceChange(aType);
  if(aType == KEikDynamicLayoutVariantSwitch)
    {
    iMyScalableEngine->Rescale();
    SetExtentToWholeScreen();
    }
  }
```

Detecting a Layout Change on UIQ 3

Again UIQ deals with layout changes as screen configuration changes rather than a simple landscape/portrait flag. An application is notified when the screen configuration changes from one of the standard screen configurations to another. The following code illustrates how a UIQ game may deal with 'flip close' and 'flip open' configuration change events.

```
void CMyView::HandleUiConfigChangedL()
  {
  CQikViewBase::HandleUiConfigChangedL();

  TQikUiConfig config = CQUiConfigClient::Static().CurrentConfig();
  switch(config.ScreenMode())
    {
    case EQikUiConfigPortrait:
      DoFlipOpen();
```

```
    break;

  case EQikUiConfigSmallPortrait:
    DoFlipClose();
    break;

  default:
    break;
  }
}
```

3.10.6   Drawing APIs and Scalability

Table 3.5 summarizes the drawing methods described in the chapter and how they deal with scalability and orientation.

Table 3.5 The scalability of drawing methods on Symbian OS

Graphics/UI type	Scalability	Notes
Drawing via WSERV	Manual	Drawing using a window server graphics context can be made scalable and will work with rotated screen modes.
Drawing via WSERV – off-screen bitmap	Manual	It's possible to create an off-screen bitmap of the same size as the screen, or one of fixed size. However, it's usually very inefficient to scale a bitmap which is not a multiple of the screen size.
Direct Screen Access + CFbsScreenDevice	Manual	Bypasses WSERV, but still allows the graphics context to be used for drawing. Code which scales using a standard graphics context can scale with this scheme also.
Direct Screen Access + frame buffer	Manual	A lot of effort is required to support multiple screen depths and rotation.

Table 3.5 (*continued*)

Graphics/UI type	Scalability	Notes
Anti-tearing API (`CDirectScreenBitmap`)	Manual	Provides a fast way of drawing, synchronized with the refresh rate of the screen, but it also represents a significant effort if support for different screen sizes, depths and orientations are required.
OpenVG/Flash and OpenGL ES	Automatic	Used for 2D/3D vector graphics.

3.11 Rendering Text

Symbian OS is internationalized, and it provides a fairly mature text rendering subsystem which supports Unicode text and bi-directional rendering. Device manufacturers usually provide plug-ins in order to render scalable fonts depending on the region in which a handset is deployed (for example Europe, the Middle East, or Asia Pacific).

However, the heritage of games means that it's common to use custom fonts tailored to the look and feel of the game. Rarely does a standard Times New Roman font fit in with the fun and aesthetics of a game, so a graphic designer typically provides a bitmap font.

There are two common ways of rendering text in games, which are as follows:

- using the Symbian text drawing services which support arbitrary scalable fonts, Unicode text, and bi-directional rendering

- copying pre-drawn characters from an image created by a graphics designer.

The first option is the most scalable across locales and market segments. In addition, since scalable fonts can be installed onto a device after it has left the factory, it's possible for a game's SIS file to include the installation of a custom font in order to differentiate the game. However, rendering large amounts of text in this way may have an impact on performance cost and variability, particularly if characters are rendered into a cache and the cost of re-rendering a character is significant.

The second option used to be very common in games. The graphics designer creates a character set within an image by creating a grid and constraining the character drawings to that grid. A character is rendered by calculating the grid reference of the character and blitting the fixedsized rectangle to the correct destination.

Figure 3.19 shows part of a fixed-size font stored as a large wide bitmap. section 3.9.3 shows some example code for extracting parts of a bitmap.

Figure 3.19   Selecting the character 'F' from a hand drawn font

Using pre-drawn characters allows graphics designers to have ultimate control of how a font looks on screen, and effects such as colors, shadows, outlines, and texture can easily be applied. However, this technique may not be scalable for widely localized and distributed games, since each possible character must be hand drawn. Some of the system fonts built into Chinese language phones have thousands of characters.[11]

The choice of methods is usually made by comparing the effort and target market with the resulting aesthetics. If a game will be localized to many languages, then it's not practical to supply a bitmap for each glyph.

The next section details the text rendering facilities provided by Symbian OS, which deal with the rendering of Unicode text.

3.11.1   Font and Text Rendering

Text-rendering API Overview	
Library to link against	`cone.lib`
Header to include	`coetextdrawer.h`
Required platform security capabilities	None
Key classes	`CCoeTextDrawerBase,` `XCoeTextDrawer`

---

[11] For a good introduction to this subject, please see *Characters: A Brief Introduction* at ***www.bellevuelinux.org/character.html***.

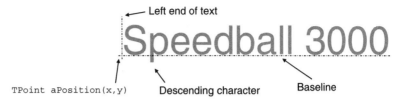

Figure 3.20 Rendering text from a coordinate point

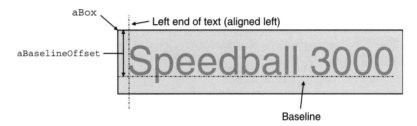

Figure 3.21 Rendering text within a box

A graphics context provides two text drawing primitives to allow you to:

- draw text from a coordinate point, shown in Figure 3.20
  This primitive draws the text from the coordinate point toward the right, justified according to the graphics context settings. The x coordinate aligns with the left side of the first character; the y coordinate with the baseline of the string.

- draw text within a box, shown in Figure 3.21
  This primitive draws text within a box, justified and aligned according to the graphics context settings and parameters to the function call, and clipped to the box. Any other area within the box is painted with the brush color.

The text is drawn in the pen color and with drawing mode set up in the graphics context, but pen style and pen width are ignored.

The code fragment below demonstrates rendering based on the x,y position.

```
// In this example, we use one of the standard font styles
const CFont& font = ScreenFont(TCoeFont::LegendFont);
gc.UseFont(fontUsed);
gc.SetPenColor(KRgbBlack);

TPoint pos(50,50);
_LIT(KGameText,"Speedball 3000");
gc.DrawText(KGameText, pos);
```

Drawing text within a box is very similar, but a bounding rectangle must be supplied with an offset and an alignment.

```
// Draw some text left justified in a box,
// Offset so text is just inside top of box

TRect box(20,20,250,100);
const TInt baseline = box.Height() /2 + fontUsed->AscentInPixels()/2;

const TInt margin=10; // left margin is ten pixels

gc.SetBrushStyle(CGraphicsContext::ESolidBrush);
gc.SetBrushColor(KRgbDarkGray);
gc.SetPenColor(KRgbWhite);

_LIT(KGameText,"Speedball 3000");
gc.DrawText(KGameText,
           box,
           baseline,
           CGraphicsContext::ELeft,
           margin);
```

3.11.2   Bi-directional Text

To support left-to-right writing languages (common in Europe) as well as right-to-left languages (for example, Arabic and Hebrew) controls that draw text to the screen can make use of the bi-directional rendering support provided by Symbian OS. When targeting multiple world regions, the use of the `DrawText()` method of `CGraphicsContext` and derived classes is now strongly discouraged in favor of the following approach.

When drawing text in scripts that run from right-to-left, much more text preparation is required than when drawing left-to-right scripts like English. These preparations include reordering the characters in the text from logical left-to-right to visual right-to-left order, as well as finding the right forms for the characters. For example, in Arabic, the character form depends on the character that follows.

To support bi-directional text presentation, controls that draw text on the screen itself store the text in a `TBidiText` object, and the `XCoe-TextDrawer` class is used to render to a graphics context. `TBidiText` objects are created and modified outside the drawing function. The content would usually be loaded from a localizable resource file so that multiple languages can be supported in a single game binary.

The following fragment shows how to set up two lines of text:

```
_LIT(KTextToRender, "Speedball\n3000");
TBidiText* iText = TBidiText::NewL(KTextToRender,2);
```

The `TBidiText` class provides methods for setting the number of lines, line breaking characters and rendering direction of the text. Assuming iText is a TBidiText object, the code below demonstrates how to render the text using one of the stock fonts available.

```
void CMyControl::Draw(const TRect& aRect)
  {
  const CCoeFontProvider& fontProvider = FindFontProvider();
  const CFont& font = fontProvider.Font
      (TCoeFont::LegendFont(), AccumulatedZoom());

  XCoeTextDrawer textDrawer = TextDrawer();
  textDrawer->SetAlignment(EHCenterVCenter);
  textDrawer.DrawText(gc, iText, rect, font);
  }
```

A key thing to note is that, unlike on previous versions of Symbian OS, on Symbian OS v9, a control can no longer keep a pointer or reference to a CFont object because the zoom factor of the control may change, for instance, if the screen mode changes.

Detailed descriptions of font selection and rendering can be found in the S60 3rd Edition and UIQ 3 SDK documentation.

3.12 Playing Video Clips

Full motion video has been a staple of console and PC games for a long time, but is not that common in current mobile games because storage space and network costs have been at a premium. However, some mobile handsets are now being equipped with 8 GB persistent storage (for example, the Nokia N95 8 GB or the Nokia N81 8 GB) and others are accepting similarly large memory cards. It now becomes feasible to construct games which contain video introductions and cut away sequences ('cut scenes').

The under-the-hood mechanics of video playing on Symbian smartphones is quite complex and may vary across devices. Highly compressed video files often require a large amount of processing to display at an acceptable frame rate, and so, silicon vendors have stepped in to incorporate IP blocks and co-processors specifically designed for offloading repetitive video codec operations.

Video codecs which are decoded on the CPU are sometimes called 'soft codecs' whereas codecs running on a co-processor may be referred to as 'hardware accelerated' codecs (even though these are written in DSP code which is also software). It's not always true that a hardware accelerated codec is faster than a soft codec, because it very much depends on the speed of the CPU, the co-processor, and the quality of the codec. The chief advantage of using hardware acceleration is that the CPU can do other things in parallel while not burdened by the codec task. Hardware accelerators may also provide better energy efficiency if they consume less power than having the CPU running at full speed during audio/video playback.

From an API point of view, Symbian OS abstracts video playback with the Video Player Utility, which we'll discuss next.

3.12.1 `CVideoPlayerUtility`

Video Player Utility API Overview	
Library to link against	`mediaclientvideo.lib`
Header to include	`videoplayer.h`
Required platform security capabilities	`MultimediaDD` (for setting priority)
Key classes	`CVideoPlayerUtility`, `MVideoPlayerUtility Observer`

As Aleks describes in the next chapter, the Symbian multimedia framework (MMF) provides a fairly straightforward utility for playing audio clips; it can also be used for rendering video clips to the screen. The video player framework can play back formats via MMF plug-ins and, as with other MMF plug-ins, they are sourced and integrated by the handset manufacturer. Common video formats that are typically supported on Symbian smartphones include:

• RealVideo 8,9,10

• H.263 and MPEG-4

• H.264

You should take care to ensure that any video clips used in a game are encoded with a codec common across target phones, and also that playback is tested on a variety of phones for performance.

The class `CVideoPlayerUtility` contains a fairly exhaustive `NewL()` factory method that allows clients of the API to specify the rectangle in which the video will be rendered, by supplying the window and rectangle. The parameters passed to this method are shown in Table 3.6.

Table 3.6 The parameters required for `CVideoPlayerUtility::NewL()`

Parameter	Description
`MVideoPlayerUtilityObserver& aObserver`	A client class to receive notifications from the video player.
`TInt aPriority`	This client's relative priority. This is a value between `EMdaPriorityMin` and `EMdaPriorityMax` and represents a relative priority. A higher value indicates a more important request. This parameter is ignored for applications with no `MultimediaDD` platform security capability.
`TMdaPriorityPreference aPref`	The required behavior if a higher priority client takes over the sound output device. One of the values defined by `TMdaPriorityPreference`.
`RWsSession& aWs`	The window server session.
`CWsScreenDevice& aScreenDevice`	The software device screen.
`RWindowBase& aWindow`	The display window.
`const TRect& aScreenRect`	The dimensions of the display window in screen coordinates.
`const TRect& aClipRect`	The area of the video clip to display in the window.

The code below initiates the player with a `CCoeControl`-derived view.

```
void CGfxVideoPlayer::InitControllerL()
  {
  iPlayer = NULL;
  iPlayer = CVideoPlayerUtility::NewL(*this,EMdaPriorityNormal,
                                  EMdaPriorityPreferenceNone,
                                      iView->ClientWsSession(),
                                          iView->ScreenDevice(),
                                          iView->ClientWindow(),
                                            iView->VideoRect(),
                                            iView->VideoRect() );
```

```
// KVideoFile contains the full path and
// file name of the video to be loaded
  iPlayer->OpenFileL(KVideoFile);
  }
```

Figure 3.22 shows that the screen rectangle used to render video is specified in terms of screen coordinates. The code below shows how to use the `PositionRelativeToScreen()` method of `CCoeControl` to calculate the screen coordinates of the control.

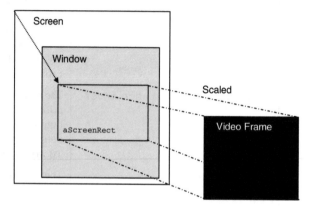

Figure 3.22 The video frame will be scaled to fit the rectangle provided (in coordinates relative to the screen)

```
void CVideoPlayerControl::ConstructL(const TRect& aRect)
  {
  iPlayer = CGfxVideoPlayer::NewL(this);

  CreateWindowL();
  SetRect(aRect);
  iVideoRect = Rect();
  TPoint point = PositionRelativeToScreen();

// Rect now converted to screen coords
  iVideoRect.Move(point.iX, point.iY);
  ActivateL();
  }
```

After this call has succeeded, the supplied `MVideoPlayerUtility`-derived class will be called back through a series of events. Table 3.7 lists the events received via the callback and the required response from the game.

For the case of simple playback, the callbacks just advance the utility to the next state, until finally `Play()` can be called to start showing video frames. This is demonstrated in the code below.

Table 3.7 `MVideoPlayerUtility` callbacks

Event	Description
`MvpuoPrepareComplete(TInt aError)`	Notification to the client that the opening of the video clip has completed successfully, or otherwise. This callback occurs in response to a call to `CVideoPlayerUtility::Prepare()`. The video clip may now be played, or have any of its properties (e.g., duration) queried.
`MvpuoPlayComplete(TInt aError)`	Notification that video playback has completed. This is not called if playback is explicitly stopped by calling `Stop()`.
`MvpuoOpenComplete(TInt aError)`	Notification to the client that the opening of the video clip has completed, successfully, or otherwise.

```
void CGfxVideoPlayer ::MvpuoOpenComplete(TInt aError)
  {
  if(aError == KErrNone)
    iPlayer->Prepare();
  }

void CGfxVideoPlayer ::MvpuoPrepareComplete(TInt aError )
  {
  if(aError == KErrNone)
    iPlayer->Play();
  }
```

If an error occurs at any stage, it's up to the class to decide what to do and whether to inform the user of the failure. Typical failures will be `KErrNoMemory` or `KErrNotSupported` if there is no codec available on the device to render the video.

Figure 3.23 shows the video example mid-playback. Notice that the video has automatically been scaled to fit the application rectangle passed to it. The codec preserves the aspect ratio of the video when scaling, which results in black bars on the top and bottom. This is sometimes called 'letterboxing.'

And that's about it really, once the video has completed, `Mvpuo-PlayComplete()` will be called and the class can safely be deleted.

Figure 3.23   Example of video playback

Combining Video with Other Elements

There is currently no standard way of implementing overlays on top of video since the video rendering code may use DSA or hardware support to render frames, leaving no opportunity for an application to draw overlays on top. Video always fills the rectangle given to it, but there is no problem combining that area with other graphical elements as demonstrated in Figure 3.24.

The window that the rectangle occupies can be any size and does not have to fill the screen (as demonstrated in the mock up in Figure 3.24). It's easy to incorporate graphics around the edges of a video by using normal `CoeControl` drawing methods.

Figure 3.24   A video played with ornaments

3.13   Less Useful APIs

The following APIs may, on the surface, seem useful to the game developer but are generally not intended for that use. You have been warned!

3.13.1   Sprites

The term sprite is heavily loaded towards game primitives. A sprite in computing usually refers to a sort of overlay icon which appears over the background and which takes care of preserving the background as it moves. On Symbian OS, sprites run in the context of the window server and are only really designed for mouse pointer effects.

3.13.2   2D Accelerated Drawing Operations

If 2D hardware acceleration is implemented by a handset manufacturer, then the Symbian OS bitmap and window server code will use acceleration internally. Direct access using the graphics accelerated bitmaps functions is unlikely to work. I recommend that you use OpenGL and OpenVG instead.

3.13.3   CImageDisplay

The `CImageDisplay` class from the MMF may be useful for displaying animated GIFs and for scaling images for display on the screen, but implementations of the API are not available on S60 3rd Edition devices.

3.14   Summary

This chapter has introduced the various fundamental graphics frameworks available in Symbian OS and provided examples of how to:

- draw using WSERV and the CONE graphics frameworks
- access the screen directly using direct screen access (DSA)
- use double buffering techniques on Symbian OS
- play back video clips
- draw international text
- deal with different screen orientations
- cope with scaling for different screen resolutions.

Having got to the end of the chapter, you should now better understand the relevance of the Symbian graphics APIs to a games environment, and you can use the examples as starting points for game rendering code.

4

Adding Audio to Games on Symbian OS

Aleks Garo Pamir

This chapter covers a variety of audio technologies supported by Symbian OS and demonstrates different techniques to create sound and music on Symbian OS based devices. It presents an overview of the audio APIs relevant to game development in Symbian OS v9.1 and later. Game developers can use these APIs to play sound effects and background music in mobile games developed on Symbian OS.

4.1 Introduction

Audio has always been neglected in game development since the beginning. Developers have focused more on the graphics capabilities and less on the sound and music capabilities of video game devices (as the 'video' in video game implies). Flashy visuals, the number of colors, and the polygon counts in 3D games have always been the main attractions.

Unfortunately, this unwritten rule hasn't changed much in Symbian OS devices, even though phones have been primarily sound-processing devices from the beginning. The sound-processing functions in phones have actually been delegated to special processors inside the phones that only process data using specific voice-only codecs. Basic ringtone operations were the only audio functionality performed by the PDA portion of the phone.

Audio support in Symbian OS phones has only started to develop in recent years. One factor was the success of the ringtone business and the increasing popularity of more technically advanced (polyphonic, digitized, MP3) ringtones. In the quest for building convergence devices, the booming MP3 player market also attracted the attention of device makers. Because of the limited amount of processing power and memory available in the devices, audio support for games had been traditionally constrained to synthesized audio generation (i.e., MIDI), not unlike the first days of PC games.

4.2   Multimedia Framework (MMF)

The need to have better audio support in Symbian OS increased with the advancement of audio hardware, the availability of processors that are more powerful and the increased battery life of devices. This led to the development of the multimedia framework (MMF), which supports video recording and playback as well as audio.

The multimedia framework, shown in Figure 4.1, is an extendable framework allowing phone manufacturers and third parties to add plug-ins to provide support for more audio and video formats. For the application developer, it provides APIs that abstract away from the underlying hardware, thereby simplifying the code needed to record and play the supported formats. It also supports streaming APIs, which bypass large parts of the MMF to provide a lower-level interface that allows streaming of audio data to and from the audio hardware.

The most important part of the MMF for the game developer is the client API. This API consists of several interfaces and is used to access most audio functionality provided by a Symbian OS device. Figure 4.1 shows a representation of how the MMF client API interfaces interact with the multimedia controller framework and other components.

The MMF requires a compatible audio controller plug-in to be present on a Symbian OS device in order to support audio operations on audio data of a certain type. The controller resides in a separate thread that the MMF creates, and all the audio processing is performed in this thread. In

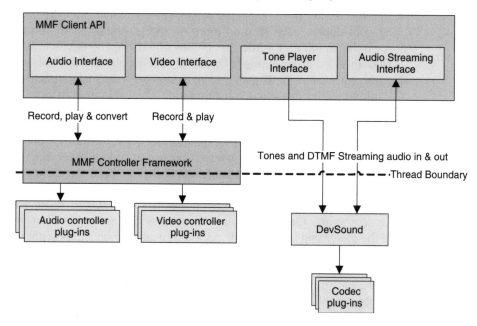

Figure 4.1   The Multimedia Framework

order to load an audio file using the MMF, the audio controller has to support the audio file format to be able to extract the audio data from the file. Moreover, in order to play the audio data, the correct audio codec plug-in needs to be present in the device.

Codec support for a particular audio type in a device is usually provided at the DevSound level as a HW device plug-in. The codecs will have low-level access to the hardware and the codec software will run on the application processor and not on a dedicated piece of hardware.

Codec support can also be provided at the MMF level as an MMF codec plug-in, but this is more of a legacy use and is no longer recommended by Symbian. The main difference between an MMF codec plug-in and a DevSound plug-in is audio streaming support. Only DevSound level plug-ins can support audio streaming.

Because of the fragmented nature of mobile devices and the multi-licensee business model of Symbian OS, it is very important for a game developer to know the technical specifications of the devices targeted. For the purposes of audio development, this means knowing which Symbian OS devices support which audio codecs, so that the audio developer can create compatible audio assets for the game. It may also be required to know what specialized custom interfaces (if any) are used to initialize or configure the codec parameters because these can also vary by manufacturer.

4.2.1 Supported Codecs in Specific Devices

Although the default Symbian OS MMF implementation comes with an audio controller that supports basic audio formats (for example, WAV) and codecs (for example, PCM), most Symbian OS phones provide support for playback of many popular formats such as MP3, advanced audio coding (AAC), and AMR. Phone manufacturers supply additional controller plug-ins only to selected devices because of dependencies on specific accelerated hardware components, licensing issues, DRM requirements, or other business factors. Manufacturers select audio support as one of the differentiating factors between the devices of different licensees. There are different levels of support even between different devices from the same manufacturer, which causes audio to be one of the areas most affected by device fragmentation.

Usually these advanced formats are more useful to game developers because they are compressed and usually have hardware accelerated decoders. The reduced audio file sizes and increased processor time for other game engine tasks are a requirement for advanced games.

Although it is possible to have audio plug-ins for a particular codec installed with the game itself, these codecs will be software-only and will not be able to use accelerated hardware. Therefore, in order to select the most appropriate audio format for a game project's audio assets, the developer needs to evaluate the available options.

By default, Symbian OS provides the 'Symbian Audio Controller' to support playback and recording of uncompressed pulse code modulation (PCM) data. It provides playback and recording support for PCM data contained in WAV, Au, and Raw files. It supports 8- and 16-bit PCM formats and IMA-ADPCM (Adaptive Differential PCM), Mu-law, A-law, and GSM 610 PCM variants.

On top of the default codecs provided by Symbian, both S60 and UIQ phones have their own sets of supported codecs, as shown in Table 4.1.[1]

Table 4.1 Codecs supported in some current S60 and UIQ devices

Platform	S60 3rd Edition	S60 3rd Edition	S60 3rd Edition	S60 3rd Edition FP1	S60 3rd Edition FP1	UIQ 3
Phones	Nokia N71, N72, N73, N75, N80, N91, N92, E60, E61, E70	Nokia 3250, 5500, E50, E62	Nokia N93	Nokia 6290	Nokia N95	Sony Ericsson M600i, P1i, P990i, W950i, W960i
AMR-WB truetones	H/W	Software	H/W	Software	H/W	–
RealAudio 1, 7	H/W	–	–	–	–	–
RealAudio 9	–	–	–	–	–	Software
RealAudio 10	H/W	–	–	–	–	–
MP3	H/W	Software	H/W	Software	H/W	H/W
AAC – LC	H/W	Software	H/W	Software	H/W	H/W
AAC – LTP	H/W	Software	H/W	Software	H/W	–
AAC+, eAAC+	Software	Software	Software	Software	Software	H/W
WMA	Software (except E60, E61, E70)	Software (except E62)	Software	Software	Software	Software (WMA9 – only W950)
AMR-NB	H/W	Software	H/W	Software	H/W	H/W
G.711, iLBC, G729	H/W	–	H/W	–	H/W	–

[1] For up-to-date audio codec information on recent phones, check *www.forum.nokia.com/main/resources/technologies/audiovideo/av_features/FN_API_audio_codec_tables.html* for Nokia and *developer.sonyericsson.com/site/global/docstools/multimedia/p_multimedia.jsp* for Sony Ericsson.

A game developer should choose a format supported by both platforms in order to maintain compatibility with the widest possible range of phones. AAC-LC and MP3 are good candidates for game audio needs. They have a wide enough bit-rate support for good quality music and sound effects and they have good compression for keeping the SIS file size manageable.

4.2.2 Audio Priority and Audio Policy in Symbian OS

Prior to going into the details of the audio APIs, it is useful to point to one other important feature. Symbian OS considers audio resources as limited, and access to these resources is constrained and prioritized.

Symbian OS is an operating system designed for voice-centric devices, and audio recording and playing are the primary functions of these devices. They are extremely important for the device's main use (as a phone); also, their immediate availability is mandatory for system critical applications. The system design provides certain applications immediate access to these resources whenever they request them. The obvious example is a phone call. If the phone OS cannot get access to the speaker to play a ringtone (i.e., to 'ring' the phone), the user will not be able to answer a call (unless they're staring at the tiny phone screen all the time, which is not a desirable 'feature', as you might imagine).

As well as certain system applications having priority access to phone audio hardware, certain licensee-provided applications (for example, built-in media players) have higher priorities too. With the introduction of platform security, even third party applications can have certain priorities, provided Symbian Signed assigns multimedia capabilities to them.

Symbian OS sorts out the access rights to audio hardware and determines which application accesses the audio resource when there are conflicting requests. A multimedia policy component is responsible for making the decisions based on rules implemented in it. Each licensee might have a different set of system and multimedia applications, so they set their own set of rules.

The one rule that is important for game developers is that games are not system critical, so they will always have lower priority than other system applications. There is always the possibility that audio requests made by the game might return KErrInUse (or KErrAccessDenied in some devices), or a system application might preempt any playing audio to play an alarm sound or a ringtone.

It is important to program any audio engine considering this fact. Handling error cases properly is a good software engineering practice for any kind of application. In addition to that, there are asynchronous notifications that you can register to be informed when an unavailable resource becomes available.

The following sections will examine the different interfaces and APIs of the MMF useful for game development in further detail. The examples will be using the Symbian PCM codec (WAV format), because Symbian supplies this codec so that any phone model based on Symbian OS will support it.

4.3   Sound Effects

The most important audio task that a mobile game needs to implement is sound effects. These complement the visual feedback the user gets from the screen by providing real-time audio cues about the action going on in the game. Because of the small screen sizes, having good sound effects is even more important on a mobile device. A good explosion effect will have a bigger emotional impact on the player than a tiny animation on the small screen.

One of the major problems with supporting quality sound effects on mobile devices is the memory limitation. Good quality sound effects take up much space on the flash drive and in memory. However, with the latest advances in the storage capacities of Symbian OS devices and the increasing number of hardware accelerated compressed audio codecs, memory consumption is less of a problem nowadays.

The other remaining problem with sound effects is latency. Mobile devices have limited processing power, and non-essential audio tasks (tasks not related to the operation of phone functionality) in general are not high-priority tasks. The generation of real-time sound effects becomes very challenging unless developers give special attention to the creation of the audio engines.

In Symbian OS, multiple audio APIs give the user different levels of control over the audio system. In order to create sound effects in Symbian OS, an application can use the MMF client API or can use the DevSound interface to access the audio hardware directly. The MMF client API is easier and more straightforward to use, but control over the hardware is limited and some operations like audio streaming have a performance penalty caused by an extra layer of buffering in the framework. DevSound is the lowest level audio API available on Symbian OS. It is more powerful, but requires more knowledge of the native audio support provided by a device.

For most casual games, the MMF API will be sufficient for creating simple effects. The `CMdaAudioPlayerUtility` and `CMdaAudioTone-Utility` classes of the MMF API are used to play single audio clips or audio tones (fixed or custom non-digitized ringtones). These classes provide many methods, but only the functionality useful for generic game development tasks are in the scope of this book.

4.3.1   Audio Clip Player

For most types of games, developers use digitized sound effects stored in sound files in one of the supported formats of the target devices. The installation process transfers these files to the target device together with the game executable. Once the game starts, the game audio engine will need to play the effects in response to certain game events.

For playing digitized sound clips, Symbian OS provides CMdaAudio-PlayerUtility class, defined in MdaAudioSamplePlayer.h. This class provides the functionality to play the audio clip directly from a file on the device or from a descriptor in memory.

In order to use CMdaAudioPlayerUtility, the audio engine should implement the observer mixin class MMdaAudioPlayerCallback.

```
class CAudioDemo : public CActive, MMdaAudioPlayerCallback
    {
public:
  virtual void MapcInitComplete(TInt aError, const
          TTimeIntervalMicroSeconds &aDuration);
  virtual void MapcPlayComplete(TInt aError);
private:
  CMdaAudioPlayerUtility* iPlayerUtil;
    };
```

When the engine is constructed, it should initialize the player utility object by passing itself as the observer. The priority setting passed in the second parameter does not have any effect for applications that do not have the MultimediaDD capability. The MMF uses the priority to determine which applications will have access to the audio hardware at a given time; games as a category have a lower priority than most other applications.

For games, the last parameter 'priority preference' is more important. The default setting is EMdaPriorityPreferenceTimeAndQuality, which fails playback operations if another client with higher priority (for example, media player) is currently using the audio resource. By using EMdaPriorityPreferenceTime, it is possible to play the audio clip mixed with the current audio or have the clip audio muted.

Of course the real meaning of both of these parameters is interpreted according to the audio policy of the device, which is device manufacturer dependent, so the exact behavior will vary from device to device (or at least from manufacturer to manufacturer).

```
void CAudioDemo::ConstructL()
    {
  iPlayerUtil = CMdaAudioPlayerUtility::NewL(*this, EMdaPriorityNormal,
                                    EMdaPriorityPreferenceTime);
    }
```

The OpenFileL() method initializes the player object with the information it reads from the file header prior to playing a file. This method has overloads for a descriptor filename, an open RFile file handle or a TMMSource-based class for DRM-protected file access.

```
void CAudioDemo::OpenClipL()
  {
  _LIT(KFile, "z:\\system\\sounds\\digital\\namediallerstarttone.wav");
  iPlayerUtil->OpenFileL(KFile);
  ...
  }
```

OpenFileL() is asynchronous and it will complete by calling
MapcInitComplete().

```
void CAudioDemo::MapcInitComplete(TInt aError, const
              TTimeIntervalMicroSeconds& aDuration)
  {
  if (aError == KErrNone)
    {
    // File is ready to play
    }
  else if (aError == KErrNotSupported)
    {
    // Codec to play this file is not available on this device
    }
  else
    {
    // Other errors, handle here
    }
  }
```

After the MapcInitComplete() callback returns, the file is ready to
play using the Play() method.

```
void CAudioDemo::PlayClipL()
  {
  ...
  iPlayerUtil->Play();
  }
```

PlayFileL() will play the file asynchronously and it will complete
by calling MapcPlayComplete().

```
void CAudioDemo::MapcPlayComplete(TInt aError)
  {
  if (aError == KErrNone)
    {
    // File successfully played
    }
  else if (aError == KErrCorrupt)
    {
    // Data in the file is corrupt
    }
  else if ((aError == KErrInUse) || (aError == KErrAccessDenied))
    {
    // Another higher priority client is using the audio device
    }
```

```
else if (aError == KErrUnderflow)
  {
  // System is under heavy load, MMF could not provide the HW
  // with audio data fast enough. Best course of action is retrying
  iPlayerUtil->Play();
  }
else
  {
  // Other errors, handle here
  }
}
```

It is possible to pause or stop a clip while playing. Pause() will pause the clip, and when it is paused, calling SetPosition() can change the current playback position. A play window can be set with SetPlayWindow(), limiting the play command to play only a portion of a clip. Stop() will stop the playing clip, reset the current playback position to the start of the play window (or the clip if there's no play window), but does not close the file. Calling Play() will restart playing. However, on certain hardware platforms Pause() is implemented by calling Stop() and resume will require reinitializing and repositioning as well as a PlayData() call. You should find out the exact behavior on your chosen target devices and develop your code accordingly.

If any play command returns KErrInUse or KErrAccessDenied in MapcPlayComplete, this means the audio resource is not available at the moment. A Play() command might return this error code immediately, which means the device is in use by another client, of higher priority. MapcPlayComplete might also return the error code sometime after the clip starts playing normally. In this case, the audio clip starts playing but, prior to finishing, another higher priority client requests the audio device and preempts the currently playing audio client.

Handling this case properly is important for portability of the game because the behavior of the device in this case is licensee-dependent and the emulators behave differently from the real devices. Symbian OS provides a generic notification mechanism for audio resource availability in the MMF API through the RegisterAudioResourceNotification() method. This method registers an observer to receive notifications when the audio resources become available. The observer mixin class to implement is MMMFAudioResourceNotification-Callback().

```
class CAudioDemo : public CActive,MMMFAudioResourceNotificationCallback
  {
public:
  virtual void MarncResourceAvailable(TUid aNotificationEventId, const
                                         TDesC8& aNotificationData);
  };
```

Notification registration is available after the clip starts playing, including inside `MatoPlayComplete()`.

```
iPlayerUtil->Play();
...
TInt err = iPlayerUtil->RegisterAudioResourceNotification(*this,
                    KMMFEventCategoryAudioResourceAvailable);

if ((err != KErrAlreadyExists) && (err != KErrNone))
  {
  // Other errors, handle here
  }
```

If another audio client preempts the playing client, the audio policy server will notify this client by calling the `MarncResourceAvailable()` method. The notification data parameter will contain the position where the currently playing clip was interrupted, so the engine can resume the clip from that position using this data. This behavior is again audio policy dependent, so it will be different on different devices. The callback might not be called depending on what sort of audio interrupts playing of the clip.

```
void CAudioDemo::MarncResourceAvailable(TUid aNotificationEventId,
                            const TDesC8& aNotificationData)
  {
  if ((aNotificationEventId == KMMFEventCategoryAudioResourceAvailable)
                    && (aNotificationData.Size() >= sizeof(TInt64)))
    {
    // get the position from the notification data
    TPckgBuf<TInt64> position;
    position.Copy(aNotificationData);
    // move the current position in the clip
    iPlayerUtil->SetPosition(position());
    // resume playing
    iPlayerUtil->Play();
    }
  }
```

Unfortunately, Symbian OS does not support mixing of audio from the same application. However, audio output from different clients might be mixed under certain circumstances on some devices, if the audio policy allows.[2] The MMF does not have the concept of 'audio channels,' so simultaneous playing of different audio clips is not supported. Audio priorities and audio policies also have strict requirements from the system audio engine that complicates things.

---

[2] A notable exception to this rule is Motorola's MOTORIZR Z8 smartphone, which has an 8-channel DSP mixer that will mix multiple streams from any process/thread indiscriminately. This is more of a feature of the hardware and the audio policy server on this device than Symbian OS, so game developers should not depend on this feature in their game designs, for the time being, for portability.

Of course, in a modern game engine, it is not sufficient to play only one sound effect at a time. When background music is also considered, this limitation becomes even more severe. Symbian is aware of these limitations and is working to address them in a future release of Symbian OS.

In the meanwhile, there are still some other options. A game developer can consider two different alternative designs for implementing a multiple-effect engine with only one output channel.

One option is to implement an interrupting model where the audio engine will allow the first sound effect to play and if a second effect needs to play prior to the first one's ending, the first effect is interrupted (stopped) and the second one starts playing. The effects can have game engine-defined priorities, so the effects with higher priorities will preempt lower priority ones, but not be interrupted by them. Developers can implement such a design by using audio clip player APIs because these APIs are asynchronous and support notifications.

To store multiple effects in memory to use in such a design, multiple `CMdaAudioPlayerUtility` objects can be instantiated, and different audio clips can be loaded. The game audio engine will decide which one to play at run time. The main advantage will be having pre-initialized objects ready to be played, which will reduce latency when played in reaction to game events. The drawback of such a method is that it wastes precious MMF server and system resources (one thread will be started and some extra memory will be allocated for each player). To minimize the memory usage, `UseSharedHeap()` can be used to share the resources among multiple clients.

Another way is to load up the effects from files in descriptors and use the `OpenDesL()` method to prime them prior to playing. Because the effects will be preloaded in memory, the file load lag will not occur. Still the requirement to initialize each effect prior to playing will still introduce some lag that might not be acceptable for some types of action games. Calling `SetThreadPriority()` with a higher priority value, which can increase the priority of the sub thread MMF creates for processing the client requests, can minimize this lag.

A third alternative is to create audio clips that contain multiple effects. An accompanying data file contains the time-based indexes and durations of individual effects in the clip. The engine loads the clip file once but, prior to playing an effect, it will calculate a play window corresponding to the effect's data in the file using the time and duration data from the index file. This will cause `Play()` to play only the portion of the file that contains the desired effect.

A much better option for a multi-effect audio engine is to implement a software mixer and mix the sounds in real time prior to playing them on the only available channel. However, this requires more processing power and a continuous streaming API. Section 4.3.3 explains this option in more detail.

4.3.2 Audio Tone Player

For certain types of simple games, such as a card game or a **Tetris** clone, all the game developer wants is to provide some kind of audio feedback to the user. Therefore, realistic sound effects or full-blown audio tracks might not be required for these types of games. Sometimes, the game developer might not have the resources to spend on specialized audio content.

In these cases, just creating the simplest of effects using the synthesizer can be the best option. Granted, it does not sound great (think of first generation ringtones), but will do the job of creating the required feedback.

Symbian OS supports synthesized sound through the `CMdaAudio-ToneUtility` class, defined in `MdaAudioTonePlayer.h`. The usage of this class is very similar to `CMdaAudioPlayerUtility`. This class supports the playing of single or dual tones. The observer mixin class to implement is `MMdaAudioToneObserver`.

```
class CAudioDemo : public CActive , MMdaAudioToneObserver
  {
public:
  virtual void MatoPrepareComplete(TInt aError);
  virtual void MatoPlayComplete(TInt aError);
private:
  CMdaAudioToneUtility* iToneUtil;
  };
```

Construction is simple and requires no setup:

```
void CAudioDemo::ConstructL()
  {
  iToneUtil = CMdaAudioToneUtility::NewL(*this);
  }
```

Playing a tone requires a prepare function to be called first. For a single tone, the `PrepareToPlayTone()` function requires a frequency and duration. For a dual tone (which is comprised of two sine waves of different frequencies), the same parameters, plus an extra frequency, are required for the `PrepareToPlayDualTone()` function.

```
void CAudioDemo::PlayTone()
  {
  // for single tone of frequency 640
  iToneUtil->PrepareToPlayTone(640,
                      TTimeIntervalMicroSeconds(1000000));

  // for dual tones of frequencies 697 and 1209
  iToneUtil->PrepareToPlayDualTone(697, 1209,
                      TTimeIntervalMicroSeconds(1000000));
  ...
  }
```

The prepare functions are asynchronous as well, eventually calling `MatoPrepareComplete()`. After the callback is complete, calling the `Play()` method plays the tone. When the tone playing finishes, the MMF will call `MatoPlayComplete()` to signal the end of tone playing. The error conditions described for audio clip playing apply to tone playing as well.

Although clip and tone playing use different classes, they share a common system resource, so the multi-channel limitation still applies. It is not possible to play a clip and a tone at the same time from the same application, except when running code on Motorola's MOTORIZR Z8 smartphone, which has an 8-channel DSP mixer that will mix multiple streams from any process/thread. Since this is a feature of that particular phone hardware, game developers should not depend on the availability of this feature on other hardware.

4.3.3 Audio Output Streaming and Creating Multi-channel Sound Effects

The way to accomplish reduced latency and multi-channel audio support for a complex audio engine is by implementing a software mixer running in a dedicated high-priority thread. Audio streaming APIs are the most suitable APIs for this purpose.

Audio streaming APIs interface directly with the hardware through the DevSound interface, bypassing the MMF. Audio data will be processed directly in the applications thread when the audio streaming API or DevSound API (see section 4.3.4) is used for streaming. This is because the sub-threads created internally by the MMF for other audio operations will not be created for these cases. Therefore, games requiring high-performance must create separate threads for streaming to ensure smooth audio output.

Symbian OS supports audio output streaming through the `CMdaAudioOutputStream` class, defined in `MdaAudioOutputStream.h`. This API builds upon the same MMF architecture as other MMF APIs but has extra methods to support streaming. The `CMdaAudioOutputStream` class contains methods to feed a running stream of audio data into the DevSound interface to produce real-time audio output from the device. The observer mixin class to implement is `MMdaAudioOutputStreamCallback`.

```
class CAudioDemo : public CActive , MMdaAudioOutputStreamCallback
  {
public:
  virtual void MaoscOpenComplete(TInt aError);
  virtual void MaoscPlayComplete(TInt aError);
  virtual void MaoscBufferCopied(TInt aError, const TDesC8 &aBuffer);
private:
  CMdaAudioOutputStream* iStreamUtil;
  };
```

Construction is similar to the `CMdaAudioPlayerUtility` class. Priority and priority preference parameters are supplied to the `NewL()` factory method together with the observer.

```
void CAudioDemo::ConstructL()
    {
    iStreamUtil = CMdaAudioOutputStream::NewL(*this, EMdaPriorityMax,
                                    EMdaPriorityPreferenceTime);
    }
```

The stream has to be opened first using the `Open()` method prior to starting streaming. The `TMdaPackage*` parameter is not used in Symbian OS after v9.1, so passing `NULL` is okay.

```
void CAudioDemo::OpenStreamL()
    {
    iStreamUtil->Open(NULL);
    ...
    }
```

`Open()` is asynchronous and it will complete by calling `MaoscOpen-Complete()`. If the callback does not receive an error code, the stream is ready to use. The next step is to configure the stream for the specific data format it will process.

The `SetDataTypeL()` method sets up the data format. The definitions of the FourCC codes required for this function can be found in `MmfFourCC.h`. Not all formats are supported on all phones, so even if the platform SDK header file contains a FourCC code, it doesn't mean all the phones using that platform version will support that particular format. The DevSound API contains methods to query the list of formats supported by a certain device (see section 4.3.4).

The `SetAudioPropertiesL()` method has to be called to set the sample rate and number of channels (mono/stereo) to complete the stream setup. The `TMdaAudioDataSettings` enumeration in `audio.h` contains the values that can be used as parameters to this method.

```
void CAudioDemo::MaoscOpenComplete(TInt aError)
    {
    if (aError == KErrNone)
        {
        // setup the format of the stream data
        TRAP(aError, iStreamUtil->SetDataTypeL(KMMFFourCCCodePCM16))

        if (aError == KErrNone)
            {
            // setup the sample rate and number of channels
            TRAP(aError, iStreamUtil->SetAudioPropertiesL(
                                    TMdaAudioDataSettings::ESampleRate8000Hz,
                                    TMdaAudioDataSettings::EChannelsMono));
            }
        }
```

```
if (aError != KErrNone)
  {
  // Errors, handle here
  }
  }
```

Using the stream to play audio data involves continuously feeding it with data in the specified format. The audio data should be packaged in an 8 bit descriptor and passed to the stream using WriteL() method.

```
void CAudioDemo::PlayStreamL()
  {
  ...
  iPlayBuffer = new (ELeave) TUint8[KPlayBufSize];

  TPtr8 playBuf(iPlayBuffer,KPlayBufSize);
  playBuf.Copy(iEffectBuffer,KEffectSize);
  iStreamUtil->WriteL(playBuf);
  ...
  }
```

Once WriteL() is called, the buffer will be copied to the internal Dev-Sound buffer to be processed and played. After WriteL() returns, neither the buffered data nor the buffer itself should be modified. The framework will call MaoscBufferCopied() when data copying finishes.

```
void CAudioDemo::MaoscBufferCopied(TInt aError, const TDesC8 &aBuffer)
  {
  ...
  TPtr8 buf(const_cast<TUint8*>(aBuffer.Ptr()),KPlayBufSize);
  buf.Copy(iEffectBuffer2, KEffectSize2);
  iStreamUtil->WriteL(buf);
  ...
  }
```

This method has a buffer parameter that will actually be the buffer written to the stream when this callback is called. This same buffer should be used to send more audio data to the stream by refilling it. The copied data will start playing when this callback is called (not immediately, but after the buffers in lower levels of the audio system are exhausted – which is practically very soon), and the stream refilling in MaoscBufferCopied() should be done as quickly as possible. Otherwise, the data in the copied buffer will run out, causing an underflow that will automatically stop the streaming. The normal way to stop streaming is by calling the Stop() method. The framework calls the MaoscPlayComplete() callback when streaming ends playing. If the error code indicates an underflow but the end of the file has not been reached, it is best to supply more data by calling WriteL() to fill the stream buffer again and continue streaming.

```
void CAudioDemo::MaoscPlayComplete(TInt aError)
  {
  if (aError == KErrNone)
    {
    // playback terminated normally after a call to Stop()
    }
  else if (aError == KErrUnderflow)
    {
    // the end of the sound data has been reached or
    // data hasn't been supplied to the stream fast enough
    }
  else
    {
    // Other errors, handle here
    }
  }
```

The stream supports an internal queue, so that many buffers can be written to the stream at once. They will be played one after another in the order of queuing. Practically, only two buffers of sufficient size will be enough to have seamless audio without pauses or clicks. While one buffer is being played, the other can be filled, and this 'double buffering' method will be enough to eliminate possible audio 'flicker.'

To implement a multi-channel audio engine using streaming, a double buffering stream should be created in a separate dedicated high-priority thread and must be continuously filled with silence data (suitable for the data format being used). All the in-game sound effects should be preloaded into the memory during audio engine initialization. The game engine should process the game events and resolve which effect needs to be played at a certain moment. The main event processing thread should pass the effect ID to the software mixer thread, and the first `MaoscBufferCopied` event that runs should mix the effect data into the running stream.

The fact that buffer copying is asynchronous and non-preemptive means the playing of effects will be delayed until the last copied buffer completes playing. There is already an existing delay in the audio subsystem because of the DevSound and hardware buffers, and any extra delay might cause the game audio effects to become unresponsive. It is therefore important to choose the right buffer size while implementing the streaming audio engine.

Some compressed formats require a fixed buffer size, but any buffer size is suitable for PCM format. Actually, mixing compressed audio data is more challenging and requires more processing power, so it is not feasible to implement software mixers using compressed data formats for most types of games. PCM is the easiest to mix and fastest to process, because it is the raw format that most audio hardware uses.

A basic example implementing a software mixer can be found among the example code provided on the website for this book, ***developer.symbian.com/gamesbook***.

4.3.4 DevSound API

DevSound is the lowest level audio API available, but its implementation is hardware-dependent (that is, it differs from device to device) and is not available on all platforms. Some platforms allow direct access to DevSound through the CMMFDevSound interface. The advantage of using CMMFDevSound to implement an audio engine is that it will allow finer grained control on audio hardware than other classes and allow more control over buffer management.

The CMMFDevSound class contains methods to:

- determine which formats a particular phone supports

- initialize and configure devices

- play raw audio data and tones.

The observer mixin class to implement is MDevSoundObserver.

```
class CAudioDemo : public CActive , MDevSoundObserver
  {
public:
  virtual void InitializeComplete(TInt aError);
  virtual void ToneFinished(TInt aError);
  virtual void BufferToBeFilled(CMMFBuffer* aBuffer);
  virtual void PlayError(TInt aError);
  virtual void BufferToBeEmptied(CMMFBuffer* aBuffer);
  virtual void RecordError(TInt aError);
  virtual void ConvertError(TInt aError);
  virtual void DeviceMessage(TUid aMessageType, const TDesC8& aMsg);
private:
  CMMFDevSound* iDevSound;
  };
```

Construction is simple and requires no setup:

```
void CAudioDemo::ConstructL()
  {
  iDevSound = CMMFDevSound::NewL(*this);
  }
```

The object needs to be initialized prior to use after construction. The class supports some methods that query the hardware for supported capabilities. These methods can be called after the object is constructed, but prior to its full initialization. This way, the client can query the hardware and select a supported setup to initialize the object with the correct audio settings.

The Capabilities() method retrieves the supported audio settings in a TMMFCapabilities object. The iEncoding, iChannels and iRate members of TMMFCapabilities are bit fields. They represent the supported options as set flags from the TMMFSoundEncoding, TMMFSampleRate and TMMFMonoStereo enumerations. iBuffer-Size is the maximum size of the DevSound audio buffer. Another useful method is GetSupportedInputDataTypesL(), which returns a list of FourCCs supported on the device for playback.

The InitializeL() method initializes CMMFDevSound. The observer and the mode are passed in, together with the data type specified as a FourCC code. InitializeL() is asynchronous and needs to be called from an active scheduler loop in a state machine fashion.

```
void CAudioDemo::RunL()
  {
  // Initialize State
  iDevSound->InitializeL(*this, KMMFFourCCCodePCM16, EMMFStatePlaying);
  }
```

InitializeL() will complete by calling Initialize-Complete(). The DevSound interface is ready for configuration at this state. SetConfigL() is called to configure the playback parameters by passing in a TMMFCapabilities object containing the desired values. The playback parameters are sample rate, encoding, number of channels (mono or stereo), and buffer size.

The buffer size is fixed for some codecs; so the size set here is 'preferred' size. The bit field parameters should only contain one of the options the device supports. The encoding parameter should always be set to EMMFSoundEncoding16BitPCM, because this parameter is not used any more and the encoding is actually determined by the FourCC parameter passed to InitializeL().

Buffer size is a trade-off between a game's audio responsiveness and RAM usage; so calculating this value correctly according to the needs of the game is very important. Smaller buffer sizes give lower latency audio but will be more susceptible to underflow as data needs to be supplied much more frequently. Big buffers give more latency, but the stream will be much more robust against underflow.

For some devices, the buffer size set here determines the internal buffer size used by DevSound and setting this value properly is the only way to achieve low latency PCM playback on these devices.

```
void CAudioDemo::InitializeComplete(TInt aError)
  {
  if (aError == KErrNone)
    {
    TMMFCapabilities caps;
    caps.iRate = EMMFSampleRate8000Hz;
    caps.iEncoding = EMMFSoundEncoding16BitPCM;
    caps.iChannels = EMMFMono;
    caps.iBufferSize = 0; // Use the default buffer size
```

```
    iDevSound->SetConfigL(caps);
    }
...
}
```

Calling `PlayInitL()` method starts playback.

```
void CAudioDemo::PlaySoundL()
  {
  ...
  iDevSound->PlayInitL();
  ...
  }
```

If there is any error during initialization, the framework calls the `PlayError()` function of the observer with the appropriate error code. Otherwise, the framework calls `BufferToBeFilled()` with a buffer reference.

`BufferToBeFilled()` is similar to `MaoscBufferCopied()` of `CMdaAudioOutputStream`. The audio data to be played has to be copied to the passed `CMMFBuffer` buffer in this function. The difference is that there is no automatic streaming, so the code has to call the `PlayData()` method explicitly to play the data copied into the buffer. The amount of data that should be copied to the returned buffer, or in other words the buffer size, is determined by a call to the `RequestSize()` function on the buffer.

The framework will call `BufferToBeFilled()` again when one of the internal buffers becomes available after the call to `PlayData()`. To finish the buffer-filling callback loop, the last buffer to be sent to the audio device should be marked as the last one by calling `SetLastBuffer()` on the `CMMFBuffer` object.

```
void CAudioDemo::BufferToBeFilled(CMMFBuffer* aBuffer)
  {
  CMMFDataBuffer* ptrBuf = reinterpret_cast<CMMFDataBuffer*>
                                                  (aBuffer);

  TInt playsize;
  if (iEffectSize -  iEffectMixPoint > aBuffer->RequestSize())
    {
    playsize = aBuffer->RequestSize();
    }
  else
    {
    playsize = iEffectSize - iEffectMixPoint;
    ptrBuf->SetLastBuffer(ETrue);
    }
  ptrBuf->Data().Copy(iEffect + iEffectMixPoint, playsize);
  iEffectMixPoint += playsize;
```

```
iDevSound->PlayData();
}
```

An underflow condition will occur if PlayData() is not called soon enough after the BufferToBeFilled() callback. The framework will call PlayError() with KErrUnderflow as an error code when this happens. PlayError() will also be called with KErrUnderflow when the last buffer is played. KErrUnderflow in this context means that the playing operation has ended, either by marking the last buffer, or by not calling PlayData() quickly enough. If the underflow error is not caused by the end of data, but is a genuine underflow, the best thing to do is to restart playing.

```
void CAudioDemo::PlayError(TInt aError)
  {
  if (aError == KErrUnderflow)
    {
    // Play finished or stream not filled quickly enough
    }
  else
    {
    // Other error, handle here
    }
  }
```

Calling the Stop() method of CMMFDevSound stops the audio stream while the stream is playing. Any data that has been passed into DevSound via PlayData() but not yet played will be thrown away.

The advantage of using CMMFDevSound interface over other methods is more flexibility and more control over some parameters. It is the closest thing to what a game developer needs, in the absence of a dedicated game audio API.

4.4 Background Music

4.4.1 MIDI Music

The most common way of supporting music in games has traditionally been through MIDI music. Because MIDI only stores the notes and instruments in the music tracks, MIDI files are small and portable. Although the exact quality of the output depends on the MIDI engine and how the instruments are represented (by algorithms to generate synthesized music, or by digitized samples of the original instruments contained in sound banks), the tunes in the music tracks are recognizable on all devices that support MIDI music.

The ease of generating MIDI files is another big advantage for MIDI music. MIDI stands for 'musical instrument digital interface' and is the standard method of communication between digital musical instruments. It is used to transfer and store the instrument players' performance information, so creating a MIDI file is as easy as playing notes on a synthesizer and capturing the output.

Mobile phones first started using MIDI for polyphonic ringtones, where several notes can be played at the same time. Because of the specific requirements of mobile devices, a special MIDI specification has been developed called scaleable polyphony MIDI (SP-MIDI). SP-MIDI makes it possible to create scaleable content across devices with different polyphony. It allows the creator of the music to define channel priority order and masking, so that when the file is played on a device with less polyphony than the original track supports, only the channels marked by the composer will be suppressed. This way, the composer can create SP-MIDI files to support different levels of polyphony and can still control the downgrading of the quality in the music on lower-end devices.

Besides supporting SP-MIDI, modern mobile phones now support general MIDI (G-MIDI), mobile extensible music format (Mobile XMF) and mobile downloadable sounds (Mobile DLS), as shown in Table 4.2. With SP-MIDI and G-MIDI, the instruments assigned to a specific program are fixed. Mobile DLS files are instrument definitions containing wavetable instrument sample data and articulation information such as envelopes and loop points. It is generally used together with Mobile XMF, which is a format that contains both a SP-MIDI sequence and zero (or more) Mobile DLS instrument sets.

Table 4.2 MIDI formats supported in some current S60 and UIQ devices

MIDI/Phones	S60 3rd Ed. & S60 3rd Ed. FP1 (Nokia N Series, E Series, 6290, 3250, 5500)	UIQ 3.0 (Sony Ericsson M600, P1, P990, W950, W960)
G-MIDI	✓	✓
SP-MIDI (with DLS)	✓	✓
Mobile XMF	✓	✓
Max. Polyphony	64	40

Symbian OS supports MIDI music through the `CMidiClientUtility` class, defined in `midiclientutility.h`. The usage of this class is very similar to `CMdaAudioPlayerUtility`. This class contains methods to play MIDI sequences stored in SMF (type 0 and type 1) and XMF files as well as descriptors. It also allows playing of single notes, opening up the possibility of creating dynamic music. The observer mixin class to implement is `MMidiClientUtilityObserver`.

```
class CAudioDemo : public CActive, MMidiClientUtilityObserver
  {
public:
  virtual void MmcuoStateChanged(TMidiState aOldState,
                                 TMidiState aNewState,
                    const TTimeIntervalMicroSeconds& aTime,
                                        TInt aError);
  virtual void MmcuoTempoChanged(TInt aMicroBeatsPerMinute);
  virtual void MmcuoVolumeChanged(TInt aChannel,
                      TReal32 aVolumeInDecibels);
  virtual void MmcuoMuteChanged(TInt aChannel, TBool aMuted);
  virtual void MmcuoSyncUpdate(const TTimeIntervalMicroSeconds&
                          aMicroSeconds, TInt64 aMicroBeats);
  virtual void MmcuoMetaDataEntryFound(const TInt aMetaDataEntryId,
                      const TTimeIntervalMicroSeconds& aPosition);
  virtual void MmcuoMipMessageReceived(const
            RArray<TMipMessageEntry>& aMessage);
  virtual void MmcuoPolyphonyChanged(TInt aNewPolyphony);
  virtual void MmcuoInstrumentChanged(TInt aChannel, TInt aBankId,
                                        TInt aInstrumentId);
private:
  CMidiClientUtility* iMidiUtil;
  };
```

Construction is similar to the `CMdaAudioPlayerUtility` class. Priority and priority preference parameters are supplied to the static `NewL()` factory method together with the observer.

```
void CAudioDemo::ConstructL()
  {
  iMidiUtil = CMidiClientUtility::NewL(*this, EMdaPriorityNormal,
                                    EMdaPriorityPreferenceTime);
  }
```

There are different ways to use this class to manipulate the MIDI engine. It can be used to play a MIDI file, to play individual notes or to do both at the same time. To play a MIDI file by itself, a call to `OpenFile()` method to open the file is required as the first step.

```
void CAudioDemo::OpenMidiL()
  {
  _LIT(KFile, "c:\\private\\E482B27E\\empire.mid");

  iMidiUtil->OpenFile(KFile);
  ...
  }
```

The MIDI engine is a state machine running in an active scheduler loop, and `OpenFile()` will kick off a state transition of the engine from closed to open state. All state transitions will complete by calling `MmcuoStateChanged()` and pass the old and the new states. By using these two state variables, it is possible to handle the state transitions.

```
void CAudioDemo::MmcuoStateChanged(TMidiState aOldState,
                                   TMidiState aNewState,
    const TTimeIntervalMicroSeconds& aTime, TInt aError)
  {
  if (aError == KErrNone)
    {
    if ((aOldState == EClosed)&&(aNewState == EOpen))
      {
      // open finished
      }
    else if ((aOldState == EPlaying)&&(aNewState == EOpenEngaged))
      {
      // play finished
      }
    ...
    }
  else
    {
    // Other error, handle here
    }
  }
```

After `OpenFile()` is called, the engine will be in `EOpen` state, ready for playing. Calling `Play()` method will start playing the MIDI file.

```
void CAudioDemo::PlayMidiL()
  {
  ...
  iMidiUtil->Play();
  ...
  }
```

During the playing of the MIDI file, the framework will call other observer callbacks, like tempo or volume change, as well. Calling `Stop()` stops playing of the MIDI file.

```
void CAudioDemo::PlayMidiL()
  {
  ...
  TTimeIntervalMicroSeconds fadeOutDuration(2000000);
  iMidiUtil->Stop(fadeOutDuration);
  ...
  }
```

One important thing to note here is that the end state of the MIDI engine will differ according to the state the engine was in when the method was called. This is also true for other state-changing methods like `Open()` and `Play()`. Therefore, it is necessary to detect state transitions in `MmcuoStateChanged()` rather than relying only on the end state.

The second way of using this engine is its capability of processing individual MIDI messages to generate audio on the fly. This is especially useful when combined with the MIDI resource playing capability.

Generating MIDI messages in real time allows the triggering of sound effects on one or more MIDI channels set aside for that purpose, while the background music is playing on a separate channel, effectively using the MIDI engine as a software mixer.

One other potential use of this capability is the ability to change certain engine parameters on the fly, which opens up the possibility of creating dynamic music. It is possible to have a single music composition that can be played with one set of instruments and a specific tempo to create a 'sad' tune, yet played with a different set of instruments and a different tempo to create a 'happy' tune. With the ability to change tempo, instruments, and other parameters on the fly using the CMidiClientUtility API, it is possible to make dynamic music transitions in response to in-game events.

The simplest way to use the MIDI engine in real time is to play individual notes by calling the PlayNoteL() method. The MIDI engine must be in a state where it is processing MIDI events (EOpenEngaged, EOpenPlaying or EClosedEngaged) prior to sending the notes. There are other prerequisites as well. The channel on which the note will be played needs to have an instrument, and the channel volume (as well as the general volume of the engine) must be set to an audible level. Otherwise, even if the function call succeeds, no sound will be heard.

```
TInt maxVol;
maxVol = iMidiUtil->MaxVolumeL();
iMidiUtil->SetVolumeL(maxVol); // overall volume of the MIDI engine
                                               set to maximum

iMidiUtil->Play(); // MIDI engine starts processing events
...
// Wait for MIDI engine to transition to an event processing state
// in MmcuoStateChanged
...
iMidiUtil->SetInstrumentL(1, 0x3C80, 17); // Instrument 17 (Drawbar
                           Organ) in bank 0x3C80 is set on channel 1

// Maximum channel volume in 32-bit real format
TReal32 maxVol = iMidiUtil->MaxChannelVolumeL();
// Set volume of channel 1 to maximum
iMidiUtil->SetChannelVolumeL(1, maxVol);
...
// Wait for MIDI engine to receive the callback to MmcuoVolumeChanged
// to make sure MIDI engine processed volume change event
...
TTimeIntervalMicroSeconds duration(1000000);
iMidiUtil->PlayNoteL(1, 60, duration, 64, 64); //play middle C on
// channel 1 for 1 second at average velocity
```

Calling the Close() method is required for cleanup when the MIDI engine is no longer needed (for example, if there is no music mode in a game, or during application exit prior to destruction).

```
void CAudioDemo::PlayMidiL()
  {
  ...
  iMidiUtil->Close();
  }
```

The `CMidiClientUtility` class supports a wealth of methods for MIDI event and message processing (`SendMessageL()`, `SendMipMessageL()`, `CustomCommandAsync()`, `CustomCommandSyncL()`), custom bank manipulation (`LoadCustomBankL()`, `LoadCustomInstrumentL()`, descriptor versions, associated query and unload methods), play control (`SetPitchTranspositionL()`, `SetPlaybackRateL()`, `SetPositionMicroBeatsL()`), and many more.

The details of this class are beyond the scope of this book, but game developers are strongly encouraged to read more about this class in the Symbian OS Library, which is part of the SDK, to find out how they can use it effectively in their game audio engines.

4.4.2 Digitized Music Streaming Using the Software Mixer

A more expensive alternative to MIDI for playing background game music is to play digitized music. It is more expensive because it needs more storage space for digitized audio files (even if they're compressed) and requires more CPU power to process them. Nevertheless, they sound closest to the real thing. Given the fact that Symbian OS supports playing digitized audio, it is an attractive option for rich games that have fewer storage limitations.

There are a few ways of playing digital background music in a Symbian OS game. The first method is playing the background music using audio streaming. Just like any other audio file, a file that contains the audio track of the game in a format compatible with the device can be played with the methods described in section 4.3.

Because of the single audio channel support of Symbian OS devices, if the file is played independently, it will use the only available audio channel. This will limit the game design, because the user will only be able to hear music or sound effects, not both.

One solution to this problem is to use another application (for example, the built-in media player application of the device or a special executable created just for this purpose) to play the music in the background, while the main game application itself plays the sound effects. Doing so will only produce the desired outcome on some devices where sound outputs from different processes are mixed by the policy server. This is not generic behavior and it may not even be possible to control the mixing in any way (for example, setting the volume or balance on the external application). Therefore, this solution is not recommended.

A better option is to use the software mixer approach and mix in the music, on an additional dedicated channel, together with the sound effects. The drawback is that the music and sound effects need to be encoded in a mixable (i.e., raw) format, so that they can be mixed together. Doing this rules out having compressed music because of the mixing problem. The only options are either having a short looping track or a longer but lower quality track.

In the mixing method, another trick is to stream the file from the disk while playing it. The background music file will most probably be big and it will not be feasible to load all of it into the heap. One way of achieving streaming is by implementing a buffering mechanism for file reading, similar to the buffering mechanism used for playing sounds in the MMF. When the MMF buffer requests more data, the pre-read data from the file is transferred from the file buffer to the MMF buffer. The lower quality mixing option, with file streaming, is demonstrated in the software mixer example code provided on this book's website.

One more method of mixing background music together with multiple simultaneous sound effects is to use Mobile XMF/DLS. Instead of mixing digitized effects to an audio stream inside the software mixer that is implemented in game code, game developers can create the sound effects as instruments in a custom sound bank and load them together with a MIDI file into an XMF file. The MIDI engine will take care of playing the file using the real instruments, and the game engine will send any effects to the mixer in real time as single notes of the custom effect instruments. Depending on the polyphony of the device, the MIDI engine can play many effects simultaneously with multi-track background music.

Of course, use of this method depends on the Mobile XMF/DLS support of the devices the game developer is targeting. Details of this method are outside the scope of this book, but more information is available from various sources. For example, see the MIDI Manufacturer's Association for more information on XML/DLS (**www.midi.org/about-midi/dls/abtmdls.shtml**). For authoring tools and usage related to game development, check out the companies listed at **www.midi.org/about-midi/xmf/xmfprods.shtml**.

4.4.3 Playing Digitized Music with Codecs

Playing compressed format codecs such as MP3 or AAC is no different from playing any other non-compressed format using the aforementioned MMF APIs. If the devices support the desired format, just providing the correct FourCC code to the initialization methods is sufficient to play these files using the audio clip and streaming APIs.

```
iStreamUtil->SetDataTypeL(KMMFFourCCCodeMP3);
...
iDevSound->InitializeL(*this, KMMFFourCCCodeAAC, EMMFStatePlaying);
```

The software mixing method will not work as described in the previous section, because most compressed codecs are not suitable for mixing in their encoded form. In order to mix them with audio data of other formats, they have to be decoded in real time and then mixed. Usually both decoding and mixing these formats in software will be an extremely heavy load and will starve the CPU, leaving no time for other game tasks.

Until Symbian OS provides support for mixing audio on multiple channels, using compressed codecs as in-game background music is not a feasible option. This, of course, doesn't mean that they cannot be used as background music in menu screens, demos and cut-scenes, where mixing with another audio stream effect or voice-over is not necessary.

Symbian OS also supports playing DRM-protected content using its content access framework (CAF). Certain MMF APIs have special methods that can be used to pass DRM-related information (UIDs, DRM access intent and so on) to the MMF classes that are used to access the DRM-protected audio files. Clients of the MMF APIs also need to possess the DRM platform security capability to access these files. A game executable must have this capability if it needs to access DRM-protected content on a device.

4.4.4 Extending Symbian OS with Software Codecs

Symbian OS provides third party developers with the infrastructure to extend a smartphone's audio processing capabilities. Other than using the built-in audio codecs for background music and sound effects, it is also possible to create MMF-compatible software audio codecs, and use them in a game application.

Using this method has the main advantage of making the game independent of any audio codec support from devices, if audio assets of the game are encoded with this codec. The software codec can be included in the installation package of the game and installed together with the game on any Symbian smartphone that supports the MMF.

Developing an audio engine that supports playback functionality through the existing MMF APIs has another advantage. It allows a developer to use an existing codec like PCM for development purposes. PCM is easier to manipulate and debug, which makes it more suitable for prototyping. Once the audio engine is mature and music and effects are polished, the switch to a compressed format will be quite easy. It takes only minimal changes to the engine code and a batch conversion of the audio assets from PCM to the new format.

The main disadvantage of the software codec method is the complexity of implementing the codec. Implementing a generic audio codec is a major development effort in itself. Even when there is an already developed codec (for example, open source audio codecs or proprietary codecs previously developed for other operating systems), it is still not trivial to port it to Symbian OS. The MMF software codecs are ECom plug-ins, and they have to adhere to a certain MMF compatible format.

Details of codec porting to Symbian OS are outside the scope of this book, but information is available from online sources and forthcoming Symbian Press books. The prime example for game development purposes is the Ogg Vorbis[3] codec port developed by Mathew Inwood of Symbian Ltd. This open source example allows decoding and playing back of Ogg Vorbis compressed audio files and can be downloaded from Symbian Developer Network (**developer.symbian.com/main/tools/appcode/ cpp/ogg_vorbis.jsp**). Studying this example code and accompanying documentation is a good way to start learning more about this advanced method.

4.5 Best Practice in Mobile Game Audio

Only a few years ago, mobile phone games used only simple beeps; now they have the capability for sophisticated audio. Game sounds give players meaningful feedback and communicate their progress through the game. Sounds in multiplayer games can be particularly useful to give feedback about the other players' actions. Background music is often used to set the mood of the game (particularly in the introductory screens) or while menus are visible.

However, the use of sounds in mobile games needs to be considered carefully to give a good user experience. It is important to remember that mobile games are played in different contexts. In some contexts, such as a multiplayer Bluetooth game with a friend, audio enhances the game experience, whereas in others, such as on a train or in a meeting, it can be extremely annoying for the player or other people! A game should be designed so it is easy to switch the context of the game, to play it silently or to play it with all sounds (music and sound effects) enabled. This is usually done through a menu option, giving the player full control over the volume. However, the game's audio should also comply with the profile of the phone, so if the user switches it into silent mode, the game should automatically become silent too.

Another issue to consider with mobile games is that they may be played in environments where it is difficult to hear well over background noises. The sounds used must be easily distinguishable from each other in the game and should also be sufficiently different to common ring tones and alert tones, the use of which could otherwise confuse the player.

---

[3] Ogg Vorbis is an open source multimedia format used by major game developers on other platforms. See **www.vorbis.com** for more information.

Sounds should not be unpleasantly high pitched, and the default setting for each should not be unpleasantly loud.

4.6 What's Next?

Traditionally, developing cutting-edge games relies heavily on the maximum use of the available hardware and OS resources. In the intensely competitive and innovative field of mobile phones, new devices with a series of new hardware and software components come to market almost every other week. It is imperative for a game developer to keep up to date with available technologies.

Symbian OS is a common code base shared by multiple licensees who use it in their devices. Each licensee adds the necessary supporting software to its hardware on top of Symbian OS. In order to use the features specific to a device, licensees provide their own APIs. If these devices are successful, the 'latest' features on these devices become commonplace in time, and all the phones on the market support this feature by default. Think of 'camera phones' or 'MP3 phones,' which started as a 'features' but are now supported by almost all smartphones, and all but the most basic feature phones.

At this point, supporting multiple APIs in their respective SDKs to do essentially the same thing does not make sense for the licensees. Instead, these different licensee APIs are consolidated into a common set of APIs supported by Symbian OS, essentially becoming a part of the shared code base. The consolidation process creates new Symbian APIs that either closely resemble licensee APIs, or that adopt an industry standard.

The Internet is the best source of information on the available technologies, closely followed by books. Unfortunately, most books about software development topics suffer from a common problem. The information they contain becomes obsolete fast because of the rapid changes in the topics they describe. This is less of a problem in theoretical books but more evident in books trying to explain the features in a specific version of a software package.

The following sub-sections attempt to overcome this problem by providing some information on the upcoming audio standards and some audio APIs supported by specific licensees. Although Symbian OS does not support these APIs at the time of writing, they are a good representation of the areas where Symbian will focus its audio support in the future.

4.6.1 Audio Effects API

This API is specific to S60 3rd Edition, and is an audio API for adding generic effects to a rendered audio stream. This API is an optional feature and S60 does not support it on every device. Even if it is supported, full support of the API is not guaranteed, and only some features might be supported on a particular device. The API is extendable, so new custom effects can be implemented.

The API consists of an abstract effect class `CAudioEffect`, an observer mixin class to receive effect-related events, and a number of concrete effect implementations.

The effects currently supported are bass boost (`CBassBoost`), equalizer (`CAudioEqualizer`), loudness control (`CLoudness`), reverb (`CEnvironmentalReverb`) and stereo widening (`CStereoWidening`).

For equalizer, reverb and stereo widening effects, there are other APIs to access preset values for these effects on the device. `CAudioEqualizerUtility`, `CEnvironmentalReverbUtility` and `CStereoWideningUtility` contain functions to access and modify the presets stored on the device.

4.6.2 3D Audio

3D audio is an area that will become more important with the transition from 2D to 3D graphics in mobile games. 3D audio support becomes even more crucial to set the mood of a 3D game on a platform where devices have a limited display size by definition.

Currently 3D audio support is included in the S60 3rd Edition SDK as a subset of the audio effects API. Doppler (`CDoppler`) and distance attenuation (`CDistanceAttenuation`) effects are two 3D audio effects provided by `CAudioEffect`-based classes in the SDK.

There are two other abstract audio effect classes, `CLocation` and `COrientation`. `CLocation` defines the methods to store and access a 3D coordinate (both spherical and Cartesian coordinates are supported), and `COrientation` defines an orientation vector. The concrete classes derived from these two base classes define the location and the orientation of a listener (`CListenerLocation`, `CListenerOrientation`) and a sound source (`CSourceLocation`, `CSourceOrientation`).

When the game loop updates the positions of the objects in a 3D game world, it must update both the 3D graphics engine and the 3D audio engine together to keep the game world in an audio-visually consistent state. By using these classes, a game developer can set the 3D locations and orientations of game objects (including the player) in the game world.

Please refer to the S60 developer documentation for more information on audio effects and 3D audio APIs.

4.6.3   Beatnik MobileBAE

MobileBAE is a commercial cross-platform audio engine developed by Beatnik Inc. (***www.beatnik.com***). It started as a software wavetable synthesis engine for playing MIDI music, but has added support for playing some digital audio formats and other advanced audio features as well.

Many licensees licensed the Symbian OS port of this engine and used it in a number of Symbian smartphones to provide the implementation of the audio features. The implementation comes in the form of standard MMF plug-ins and is accessed through Symbian Client APIs. It also provides C and C++ APIs for its native functionality, but this API is not exposed to the third party applications. Sony Ericsson provided an extension to their SDKs for some Symbian OS devices that used an older version of this API, but UIQ 3 devices use the new version.

Please contact Beatnik Inc. directly for more information about this engine.

4.6.4 Helix DNA Client for Symbian OS

Helix is a cross-platform digital media framework originally developed by RealNetworks and was subsequently released to the public as an open source project. It consists of a media server, a client, a player application, a codec and format framework, and a content generation application, called 'the producer.' This framework can be used to stream and play video and audio data in various formats.

Helix has a Symbian OS port of the DNA Client that is used to build player applications compatible with various versions of Symbian OS including v9.1. Multiple Symbian OS devices have been built that used Helix technology for their own media playing features. The Helix client and player projects are open source and are under active development. The current version supports audio codecs including MP3, AMR-NB, AAC+, RA8-LBR, RA10, RA-Voice, WMA9, WMV9 and Vorbis, as well as various formats that can contain these types of data.

The Helix source code is a good example of audio software. It can be used in open source projects as well as being licensed for commercial projects. It is also possible to use the Helix Client as a middleware library on devices where it is included. For more information on Helix, please visit the Helix Community website at *helixcommunity.org*. You can find Symbian OS-related projects at *symbian.helixcommunity.org*.

4.6.5 Khronos Standards

The Khronos Group (*www.khronos.org*) is a member-funded industry consortium focused on the creation of open standard APIs to enable the authoring and playback of dynamic media on a wide variety of platforms and devices. Symbian is a member of the group, as are a number of Symbian licensees and partners.

The Khronos standards for audio are OpenMAX AL and Open SL ES (described below). As we go to press, the Khronos Group has just announced that provisional versions of the OpenMAX AL 1.0 and OpenSL

ES 1.0 specifications are available.[4] Khronos expects the specifications to be finalized by mid-2008 after integration of industry feedback and completion of tests to enable conformant implementations to use Khronos trademarks.

Other Khronos standards of interest to mobile game developers, for example for graphics or hardware interaction, are described in Chapter 7.

OpenMAX

OpenMAX AL is a media library portability standard. It defines a royalty-free cross-platform API that provides portability for streaming media codecs and applications.

This standard tries to reduce fragmentation of codecs, multimedia frameworks, and multimedia applications by providing three levels of abstraction. These three layers can be implemented together or independently from each other. Each software component supporting this standard has to provide APIs that conform to the API specifications defined by this standard.

The lowest layer of abstraction is the 'development layer' (or OpenMAX DL) that defines media primitives and concurrency constructs. Hardware media engines usually expose this layer. The intermediate layer is the 'integration layer' (OpenMAX IL) which defines media component interfaces. This is an OS media framework layer. The highest level of abstraction is the 'application level' (OpenMAX AL) which defines high-level playback and recording interface APIs. This is the level used by applications running on a mobile OS.

Symbian is currently implementing support for OpenMAX IL compliance to make multimedia system integration easier for hardware device creators and to improve the audio adaptation framework.

The most relevant API set for game audio developers is the application level OpenMAX AL. It provides a standard high-level multimedia API for audio playback and recording as well as video, image capture, and radio control. It consists of a set of objects (for example, engine, devices, media objects) and interfaces (for example, play, seek, record, audio encoder, radio, MIDI). Although definition of the API is nearing completion, it is not anticipated to be available on Symbian smartphones in the near future.

---

[4] *Khronos Group Releases OpenMAX AL 1.0 and OpenSL ES 1.0 Specifications for Embedded Media and Audio Processing*, October 2nd, 2007: **www.khronos.org/news/ press/releases/khronos_group_releases_openmax_al_10_and_opensl_es_10_ specifications_for_em**.

Open SL ES

OpenSL ES is another standard from Khronos Group. It is designed to be the sister standard of OpenGL ES in the audio area. It is designed to minimize fragmentation of audio APIs between proprietary implementations and to provide a standard way to access audio hardware acceleration for the application developers. OpenSL ES is also a royalty-free open API and is portable between platforms, like OpenMAX. In fact, OpenMAX AL and OpenSL ES overlap in their support for audio playback, recording and MIDI functionality. A device can choose to support both OpenMAX AL and OpenSL ES together or just one of them.

OpenSL ES supports a large set of features, but these features have been grouped into three 'profiles': phone, music and game. Any device can choose to support one or more of these profiles depending on its target market. OpenSL ES also supports vendor-specific extensions to add more functionality to the standard feature set of a profile.

OpenSL ES Game Profile supports hardware accelerated playback of audio files, advanced MIDI functionality (SP-MIDI, Mobile DLS, Mobile XMF, MIDI messages), audio effects (equalizer, reverb, stereo widening, Doppler, pitch control), audio buffer queues, 3D audio, and more.

OpenSL ES API is identical to OpenMAX AL, but adds support for more objects (for example, listener, player and 3D Groups) and interfaces (for example, 3DLocation, 3DPlayer, 3DDoppler, 3DMacroscopic) for 3D audio support.

5

Multiplayer Games

Leon Clarke (Ideaworks3D)
Michael Coffey and Jo Stichbury

5.1 Introduction

Mobile phones are networked by nature – they are, after all, designed to be used to communicate and engage with others – and are therefore well suited for multiplayer games.

In this chapter, we will discuss aspects of game design on Symbian OS for players who are local to one another (local multiplayer games using short link connections) and for players who are geographically distant (networked, or online, multiplayer games). In either case, the number of players of a multiplayer game is, as the name suggests, more than one. However, it is not limited to two. Depending on the game's design, a number of people can play a mobile game together – either as a team (or two teams, playing against each other) or as opposing individuals.

In a single-player game, the player plays against the game, and is challenged to either complete an action at a given time – or as quickly as possible – or solve a mental puzzle, or some combination thereof. The player is challenged by the artificial intelligence (AI) of the game alone. In a multiplayer game, the player is still challenged by the game, but also interacts with others, either playing against them or working as a team against other players or the game's AI.

5.1.1 Connection Drops

In a multiplayer game, a player may leave the game before it is finished, perhaps by choice[1] (for example, if the player is interrupted and needs to

---

[1] A game designer should avoid giving players an incentive to leave a game. If a game has a leader board where their rank improves if they win, and declines if they lose, but is unaffected by dropping out of a game, they are likely to leave a game they are pessimistic about winning.

do something else) or because of a dropped connection (for example, if a train goes into a tunnel).

Once a player's departure is detected, the game must handle it gracefully. If there is only one player left, the game may stop and declare the remaining player the victor. If there are still multiple players remaining, the game designer can opt to keep the game running so the other players are not inconvenienced. A number of techniques are feasible:

- 'Civil Disorder': the departed player remains in the game, but does nothing. This phrase was originally used in Allan Calhammer's board game Diplomacy, in which the armies and fleets of a player remain in place on the board after they have left.

- AI takeover: the game's AI takes over the leaving player's position (commonly used in online multiplayer card games). This approach works even if a single player is left after all other opponents drop out.

- Replacement player: with some types of games, it may be feasible to fill the position with a new player. For example, in online multiplayer card games, if a game is full, a player can wait to take over a position when another gamer leaves.

When a player decides to stop playing the game and do something else, with a single-player game, it's simple enough to save the game state for next time. However, with multiplayer games, it is more complex, since although one player has left, others may choose to continue playing, and certainly will not want to sit and wait indefinitely for that player to rejoin and resume where he or she left off! Some games are designed so that players may drop in and out, and the game continues regardless when they are offline. Good examples are massively multiplayer online role-playing games (MMORPGs). Other games are played to completion without the player who has departed, or are simply abandoned when the player leaves. When the player resumes the game in multiplayer mode, he or she starts a new game and does not resume the previous session, since it cannot be predicted that the player's original opponents are still online and the game in the same state as when he or she departed.

5.1.2 Latency

Latency – the time between when a device makes a request and it is received – is a big factor in multiplayer mobile games when played over a phone network. While it can be measured in milliseconds on the wired Internet and for Bluetooth connections, it can be several magnitudes higher, measurable in seconds, over the air. High latency currently makes it almost impossible to develop certain genres of mobile game, like first person shooter action games, for online multi-play over GPRS.

Although the move to third-generation (3G) technologies does expand the bandwidth available to over the air networks, it does not necessarily significantly change the latency equation. (That is, you may be able to send 100 KB in the time it used to take to send 10 KB, but it will still take a second or more for the data to start arriving.)

Game design needs to take latency into account to avoid issues of latency. Suggested approaches include:

- Turn-based games: each player takes a turn and, because the players take some time to respond, the additional seconds imposed by network latency are acceptable. However, if there are too many players, the amount of time a player must wait for a turn may be intolerable, so game designers should limit the number of players of a turn-based game accordingly.

- Simultaneous-move games: these are games in which each player decides what to do during the next turn, and submits the move, which the game sends over the network. All moves are revealed simultaneously, the winner is determined and the result sent to each player. One of the advantages of this system is that players don't have to wait through everyone else's move. A player who sends in her move quickly may have to wait until the slowest player has completed his move, but this is still faster than waiting on every turn separately. A good example of this kind of game is Roshambo – also known as Rock, Paper, Scissors.

- Solo multiplayer games: even though this sounds like a contradiction! In fact, in a solo multiplayer game, players each face the same challenge, playing in single-player mode, with scores compared at the end of the round, and a winner declared. Because information is only exchanged with the server at the beginning and end of each challenge, latency is less of an issue.

5.2 Local Multiplayer Games Over a Bluetooth Connection

Local multiplayer games are those that can be played by more than one player, each using a different phone, communicating by a short link connection. Here we will concentrate on multiplayer games over a Bluetooth connection because this is the most convenient technology for use on a Symbian smartphone.

Bluetooth is a technology that allows devices to establish a local-area wireless network. Devices must be within about 10 meters of each other to establish connections. Bluetooth connections allow for fast data transfer rates when compared to transfer over the cellular network. Over GPRS, you can expect multi-second latency, but over a Bluetooth network, latency is of the order of 20 to 50 milliseconds. The relative throughput (the amount of data flowing through a connection, disregarding protocol overhead) of a Bluetooth connection approximates 700 kbps, while comparatively, over GPRS, throughput is approximately 50 kbps. Throughput and latency both favour Bluetooth over GPRS for fast, action-based multiplayer games.

Multiplayer functionality using a Bluetooth connection can be a nice addition to a game because it is fun for users to play against friends that are in the same room without having to pay for a data connection, or have any other equipment besides their mobile handset. All S60 and UIQ smartphones support Bluetooth, and most users are familiar with its use, although, for a good multiplayer game experience, it is still important to present an easy way for players to set up the Bluetooth connection so there is no barrier to its use.

The information and example code required to provide a good multiplayer game using a Bluetooth connection is significant. We have decided that the level of detail required is beyond the scope this book. Instead, we have written a paper and created example code to illustrate how to use L2CAP sockets to create multi-point Bluetooth connections for a multiplayer game. Both the example code and paper may be downloaded from the Symbian Developer Network at **developer.symbian.com** – a link is also provided on the developer wiki page for this book (found at **developer.symbian.com/wiki/display/academy/Games+on+Symbian+OS**).

An alternative way of creating Bluetooth multipoint connections is to use the Bluetooth PAN profile to establish a local ad hoc piconet. This presents the PAN profile as a network interface, or Internet Access Point (IAP).[2] However, at the time of writing, the PAN profile is not currently widely supported on Symbian OS smartphones, so we currently have no documentation on its use in Bluetooth multiplayer games. However, we anticipate this to change, and recommend that you check the Symbian Developer Network for papers on using PAN and WLAN connections in multiplayer games.

5.3 Online Multiplayer Games

At first glance, online multiplayer games (meaning those games where communication from the handset goes over the Internet rather than over a short link technology such as Bluetooth) may seem to differ very little

---

[2] An IAP is a concept specific to Symbian OS, associated with IP-based applications. It is a collection of settings that defines how a connection to a particular network is made. An IAP can be thought of as representing a single network interface.

from short link gaming. You might initially guess that there will be some differences because the network will be slightly less reliable due to greater complexity when connecting to the global Internet.

Also, there may be differences in game design for a game that can challenge people anywhere in the world rather than just other players within Bluetooth range. There are design implications for setting up a multiplayer game session over a network with a stranger, compared to setting up a session to play over Bluetooth with someone who you can probably see and talk to as the game progresses.

However, those differences seem relatively minor. But, in fact, online multiplayer games are very different propositions to short link connected Bluetooth games for another reason. The big difference between them is that online games will almost always involve connection to a remote server. The server is needed to allow each device to know which other devices playing the same game are on the network. This is because network operators often use Network Address Translation (NAT) to prevent devices from communicating directly, even if game players do know about each other and arrange to play a game together over the network. For the purposes of the rest of this section, we always assume the presence of a remote server when we discuss online multiplayer games. Those games using an IP network (such as an ad hoc WiFi network) without a server may be simply considered a variant of short link multiplayer games, and are not discussed here.

5.3.1 Issues and Features Associated with Server-Based Online Games

Some of the issues that affect all online multiplayer games are that:

- the network operator must have a suitable data network
- the device must be set up correctly to use it
- the users understand that they may be billed for the use of data.[3]

Nowadays most phones will be configured with the correct settings for the operator, but some operators may have separate 'web' and 'data' network settings. The web settings will be billed at a lower rate but will be restricted to a 'walled garden' of content that probably doesn't include external game servers. A game developer will have problems informing

---

[3] If players are not on flat-rate ('all you can eat') data plans, they expect to be told when an application is doing things that will cost them money. But, on the other hand, the same game will be deployed through various operators to people with varying price plans, and some of those people *will* be on flat-rate plans. Those players don't want to be hassled with too many confirmation dialogues, or confused by being told that the game may cause them to see extra data charges, when it actually will not.

the user which settings they are supposed to use, since this is different for every operator.

However, once the remote server is in place, and the user can connect to it, numerous other use cases such as asset downloads, non-real-time gameplay, server-based game code, and community features become possible. These are arguably more useful and exciting than just playing against other people for the duration of the game session. Let's consider these in a little more detail.

Global real-time multiplayer gameplay

This is the most obvious application: you can play in real time against opponents anywhere in the world, just as you can with online console and PC games. The theory is simple; you have a lobby running on the server which connects you with opponents, then once the game has been set up, everything is roughly the same as with short link gaming, only over the Internet.

However, it's been used extensively in only a very few commercial mobile games on Symbian OS so far. There are a number of complications, some of which are as follows:

- Latency is high. Mobile latencies are higher than over the wired Internet or over Bluetooth, which has very low-latency, as we discussed earlier. Over the phone networks, latencies can also be rather unpredictable. Game design needs to consider this, and some genres of game will be a real challenge. Racing games can be playable with very high latencies, but other genres, such as shoot 'em ups or multiplayer platform games, are more problematic. Exact latencies vary depending on the wireless technology being used; GPRS network latencies are claimed to be in the region of 1 second but they are often higher in practice. HSDPA is expected to have latencies in the region of 100 ms, which will make far more game genres possible.

- There are various perceived and actual problems with reliability, because mobile wireless communications are inherently susceptible to all sorts of complications. Assuming you've designed a game for an HSDPA handset that you know will only be deployed on a network that's rolling out HSDPA, can your game cope when someone roams mid-game onto a GPRS network? Even worse, the device might drop out of coverage altogether.

- Flat-rate data tariffs are becoming more common but are still rare in many areas. The cost of playing online games may be high, and it is also hard to explain to the user, since the game has no way of knowing what price plan the user is on. Game publishers are understandably very wary of giving users nasty surprises in their phone bills. Even network operators would prefer to avoid users running up high bills unexpectedly, since they have to pay for customer support personnel to handle the angry phone calls received in complaint!

Non-real-time multiplayer gameplay

There are other modes of multiplayer gameplay that don't involve real-time interactions. The most simple is a high score service – a player uploads their scores on a regular basis, and can see how they fit in relation to the overall population of gamers for that title.

At a greater level of complexity, game modes can be created that involve interacting with another player, but not at the same time. Many of Ideaworks3D's games for the original Nokia N-Gage game deck involved 'shadow racing' online modes. Players attempted to complete an objective in a shorter time than an opponent, by playing against the ghostly image of the other player's attempt at the same challenge, which they had previously downloaded using their GPRS connection.

Another option is to design turn-based games, either where opponents play a turn immediately after each other or with turns happening at fixed times, such as one turn per day, for example, in a prolonged game of chess or hangman.

Incremental asset downloads

So real-time gameplay is possible, but has many complications that often scare the game publishers and network operators. What can usefully be done online instead? One feature that is simple but very useful is incremental asset downloading. The current generation of smartphones have graphics capabilities that easily surpass the quality of the original Sony PlayStation, but over the air download speeds and shortage of storage space on the phones means that mobile games usually have to fit in a footprint of 1 or 2 megabytes, which is far less than the 650 MB that PlayStation game developers had to play with. An obvious way of winning some extra storage space (though still a lot less than 650 MB) is to store only some of the game on the handset at one time, for instance by incrementally downloading only a single level at a time.

A well-designed asset download system must be able to cope with the sometimes temperamental nature of mobile wireless communications. Requirements include being able to resume interrupted downloads and to keep control of the amount of storage space the game occupies.

Server-side game logic

The obvious 'next step' after level downloads is to start putting game logic on the server. For example, Ideaworks3D has developed a mobile version of a racing game where buying and customizing new cars are important aspects of game progression. For a mobile game, the title has an extensive range of cars and options because the entire car shop is implemented on the server. The server keeps track of where the player is in the game, and hence which items are available to buy, how much money the player has, and what items have been bought. Based on this information, it then assembles the player's car for the game to download. It also supplies a range of equivalent cars for players to race against to keep the game interesting. Compared to asset downloads, this is not dumb delivery of data – the server decides on the data the game should download based on knowing the player's progression through the game.

Another Ideaworks3D mobile game runs most of the rendering code from the desktop version of the game on the server, allowing the player to create highly customizable characters that can then be downloaded and used. The characters appear more detailed than would be expected on the relatively low-end phones on which this game runs, because they were rendered with desktop graphics hardware.

5.3.2   Challenges of Implementing a Game Server

So what is needed to develop a client-server game? As has been hinted above, there are a number of challenges in the mobile space; so just using a standard HTTP server (or something like that) might not work as desired. Here are a few problems that would need to be considered when designing the game server.

What protocol is used to talk to the server?

Some sort of custom protocol running over TCP or UDP gives the most flexibility. If using UDP, then reliability might need to be implemented at a higher level, but it can be turned off when reliability isn't required. Alternatively, all messages could be tunnelled over HTTP. Then, a standard HTTP server and the Symbian OS HTTP stack can take care of the protocol stuff, but the disadvantage is that, arguably, they are not very suited to the task. The message headers are relatively verbose, and all communications must fit into the model of requests from the client with responses from the server; the server cannot send unsolicited messages to the client.

How is the player (or the handset) identified?

If the connection briefly disconnects or idles out, how does the handset identify that it's the same session as last time? If using HTTP, we could have some sort of registration request which delivers a cookie used to identify the session, which is then sent with subsequent requests. Using a custom protocol, something similar probably needs to be invented.

Does the server need to be scalable?

That is, is a cluster of more than one machine necessary? And what happens in the case of failure? Here, the question of where data is stored becomes quite important. Storing a set of backup data in a database is probably a good idea, since backup, clustering, and failover of databases are well understood. Different machines in the cluster will immediately have a consistent idea about the current state of the data.

A number of other questions present themselves, including:

- How is data to be serialized for transmitting over the wire?

- Do any operations, like downloads, need to be resumed after errors? How does the client do that? Does the integrity of downloads pieced together from many attempts need to be checked?

- What expertise is needed to write this? Does your development team include anyone with network programming experience, experience of managing redundant server clusters, or database programming and administration experience?

It is clear that developing a mobile online multiplayer game is actually two development projects in one: a client-side game that runs on the handset and communicates with the server, and server-side code and management. Remote multiplayer mobile game development requires additional skill sets to those required for single-player mobile game development.[4]

Explaining all these different areas properly here would turn this book into a very heavy server networking book. What many developers need is a separate product that will deal with all the unfamiliar networking stuff and let the development team get on with writing games. The next section will look at one such product – Airplay Online from Ideaworks3D – which was written by a team that included Leon Clarke, the author of this section.

5.4   Airplay Online: A Multiplayer SDK and Service Solution

Airplay Online can be thought of as solving three problems, each of which we will look at in turn:

1. It provides a set of 'services' to perform standard operations like downloading assets, managing high scores, tracking the user's identity, and a lobby for setting up multiplayer games.

---

[4] In addition to the extra skills required, writing and maintaining an online multiplayer game has an ongoing cost, in terms of server hardware, hosting, and bandwidth costs.

2. It provides a transport protocol designed for the requirements of mobile gaming, and a server and client libraries that implement this protocol.

3. It makes it as easy as possible to extend the standard services or write new custom services, if either the standard services do things in the wrong way, or if game-specific logic is needed on the server.

As Figure 5.1 shows, Airplay Online servers run on a cluster of Linux servers, although the entire system works on Windows for development purposes (meaning that server-side code can be developed without needing access to a Linux development machine). Games can either be hosted on Ideaworks3D's cluster, or on a cluster owned and run by the game developer or publisher.

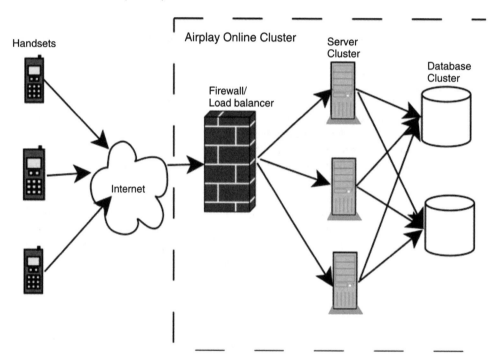

Figure 5.1 An Airplay Online server cluster

5.4.1 Standard Services

Airplay Online comes with everything that most games will need 'out of the box.' A game that just uses standard services simply needs to provide a data bundle to upload to the server and call the relevant APIs to access the services. Programmers with a fear of server-side coding can write a multiplayer game with no problems. The range of standard services is growing, but currently includes a login service, a download service, a high score service, and a lobby service.

The login service associates the player's online identity with the IMEI of the phone. This login happens automatically when the user first connects to the server. This also keeps track of a user's tag/name if the game needs players to have names (for example, to identify players on the high score service). Apart from anything else, player identification is important because it enables publishers to know how many people are playing the game.

The download service resumes downloads that have been interrupted and uses digital signatures to check the validity of downloaded files. This is combined with a utility library for keeping track of a game's use of storage space, which works by claiming all the space the game will ever need when it is first run. Before downloading a new file, the utility will decide which (if any) files need to be deleted to make space for it, and create a file of the right size. The download service then over-writes it with the data. In this way, the game doesn't unexpectedly run out of disk space.

The high score service allows games to create any number of different tables (for example, to represent different levels). As well as simply storing the scores and who scored them, the high score service can store a buffer of information with each score, which can be used to provide information needed to re-play the record-breaking runs, or other general information about how the score was achieved.

The lobby service is used for the setting up of multiplayer games, by creating challenges, letting other players search for created challenges and joining them, and then starting off the game, setting up the communication between the players and synchronizing start times.

5.4.2   The Transport Layer and Basic Server Infrastructure

As far as is possible, the transport layer is designed so that it 'just works.' In simple usage, the game developer can be almost unaware of it. The game simply needs to create a session with the server at start up, then start using the services. Errors may get propagated up via the services, and that's about it.

Figure 5.2 shows the typical sequence of performing operations using the polling style of the API. The general style of interaction (e.g., connect, do things, poll for the status of the thing you've done) is the same for all services and operations on the service. There is, in fact, an alternative style of API which uses call backs instead of polling, if the game developer prefers it.

However, when doing slightly more advanced things, more facilities are needed. Airplay Online attempts to 'make everything as simple as possible, but not simpler' (to quote Einstein) by providing increasing

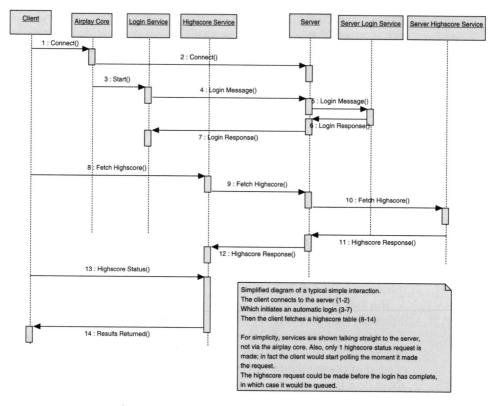

Figure 5.2   A typical Airplay Online interaction sequence

layers of complexity that can be used, and worried about, if the game demands them.

All traffic is multiplexed over one socket, to make life easier when considering firewall setups and when explaining what ports the game uses. It will run over either TCP or UDP. For everything except real-time multiplayer, TCP is probably best. The protocol implements reliability on top of UDP, which used to have the major advantage that it worked round various eccentricities in different TCP implementations, but nowadays all the major mobile operating system vendors have properly working TCP stacks. For real-time multiplayer games, running over UDP has the major advantage that the protocol can turn off reliability for any given packet. You might think that reliability would inherently be a good thing, but if the data is real-time, it is usually better to send new, up-to-date, data than to spend a long time attempting to re-send out of date information.

Several kinds of messages can be sent over the link. There are ordinary one-way messages in either direction, or transactions (i.e., request/response pairs). 'Backwards' transactions, where the server sends the request and the client sends the response, are also supported (although by their very nature, they only work once the client has connected to the server). Transactions are fully pipelined (more than one transaction can be happening at the same time) but, unlike HTTP pipelining, responses don't have to return in the same order that the requests are made.

Real-time packets can be flagged so that if there are two packets of the same type in any queue, the old one will be discarded. Using this feature, the game can throw real-time data at Airplay Online as fast as it likes, and Airplay Online will transmit the newest available packet as rapidly as the link (or the QoS settings) allows. Because this feature uses a separate notion of packet type to other parts of the system, a different player's data will be treated as a different type, despite the fact that each player is using the same packet format.

The Airplay Online protocol monitors the latency of the link, and uses this information to guess available bandwidth and to provide a shared clock for synchronizing game events. It can also attempt to implement a simple QoS system by sharing the available bandwidth among different services, meaning that a relatively bandwidth-intensive service, like the download service, can't swamp everything else when bandwidth briefly drops for some reason.

When a client first connects, the protocol client code on the device automatically sends quite a lot of useful information to the server, such as what game is being played, what kind of handset it is, and what language the user is using. The server-side half of the login service will add to this information with the user's player ID and tag (if defined). Services can then make use of this information, for instance to supply localized resources, to supply the correct resources for a particular game, or just to keep track of who the user is. If the session is disconnected for any reason (including time outs) then it is efficiently resumed and the server restores all the settings it had last time.

It's worth mentioning that the standard login service can be replaced by another one if required; for instance, the standard login service currently ties the identity of the player to the handset being used. If a game wanted to allow people to log on from a friend's handset, it might want to write a username and password-based login service. This custom login service can use the same mechanism that the standard login service uses for user identification.

The client-side code has various features to help the developer keep track of memory usage. Game developers like things to be very predictable so that memory allocations don't happen at unexpected times in order to ensure that the game will not run out of memory at some critical moment in the game. Keeping things predictable is a challenge, because network programming is inherently unpredictable, and incoming packets or failures to transmit outgoing packets may happen at any moment. Solutions include creating fixed-sized pools of commonly allocated structures (notably packet buffers), allowing the game to supply memory to be used for particular operations, and allowing all online allocations to be routed through a special allocator, for instance allocating on a separate heap.

The server provides a general framework for running the various services, making it easy to set up a redundant cluster with no single point of failure. Each service is run in a separate process so that games are isolated from one another. The server infrastructure generally keeps track of multi-threading, a logging framework and everything else that you'd expect from a server infrastructure. Because all data is stored in a database, managing failover and backup for the data is reduced to a standard task that will already be well understood by system administrators.

5.4.3 Custom Services

Most online game frameworks let people write custom code on the server. Airplay Online makes this as easy as possible, instead of making sure that the standard services do everything anyone could ever want them to. This approach was chosen because even apparently 'standard' functions need to interact with custom aspects of game logic. For instance, a game might have some complex rules about who can play who online, based on exactly what single-player objectives have been achieved. A standard lobby can't be expected to cope with that. The Airplay Online approach is to provide the source code to the standard lobby as an example, allowing a game to write its own custom lobby that just re-writes the matchmaking code.

The key to making custom service development easy is a tool called the 'service builder' which takes an XML description of the format of messages that need to pass between the client and the server, describing the data structures in each message. From there, the service builder will build client and server framework code to pack and unpack the data, make requests, monitor the status of requests and return unpacked results. On the server, services can be implemented in either C++ or Python, with Python being preferred. Typically, writing the server component simply involves filling in a group of stub functions that have been auto-generated, each of which is passed the request data structure and provides a response data structure to fill in. On the client side, the service builder generates both the polling and the callback-based API for each message. The client code can be customized in complex cases, but it usually doesn't need to be.

The goal of this architecture is to isolate the developer from any of the complexities of network programming or server-side development. The developer is isolated from the network layer and only sees unpacked messages; on the server, developers don't need to concern themselves with the server process's creation of different threads for handling different clients and so on.

5.4.4 Summary

Online gaming allows a richer gaming experience, in ways varying from the obvious (playing against other people) to possibly less obvious ones (more and bigger textures due to incremental asset downloading). It can certainly be very popular. As we were completing this book, *Halo 3* for Xbox 360 was launched. Within the first day of its launch, *Halo 3* players worldwide racked up more than 3.6 million hours of online game playing, and 40 million hours by the end of the first week, representing more than 4500 years of continuous gameplay![5]

However much fun it may be to play online games, game programmers often perceive network programming as a difficult and unfamiliar task. This difficulty can be overcome by the use of a middleware product such as Ideaworks3D's Airplay Online, which lets people create online games with the minimum of effort. Games performing common online operations can be written with no server-side coding at all. More complex games will make the network and server side coding as simple as possible – but no simpler.

5.5 N-Gage Arena

N-Gage Arena is an end-to-end solution for online multiplayer and community services for N-Gage games running on a range of S60 smartphones. Chapter 8 discusses the N-Gage platform in more detail, but N-Gage Arena is described here because it forms a natural part of the multiplayer gameplay chapter.

N-Gage Arena was originally part of Sega.com's SEGA Network Application Package for mobile multiplayer gaming. Nokia acquired Sega.com Inc. (a subsidiary of SEGA) in early 2003 to enable them to offer networked multi-play and a virtual community to N-Gage gamers. Nokia reports almost 700,000 N-Gage Arena members in 190 countries and across 320 network operators.[6]

Players log into N-Gage Arena to play multiplayer N-Gage games over the network, and participate in other aspects of social gaming, for example, by:

- entering game tournaments and other hosted events such as celebrity interviews and live chats

- storing their player statistics (scores and standings) online, with score boards listing top-ranked players.

---

[5] More information about the *Halo 3* phenomenon can be found at ***www.microsoft.com/ presspass/press/2007/oct07/10-04Halo3FirstWeekPR.mspx***

[6] See ***www.gamesondeck.com/feature/1364ngage_arena_nokias_tomi_huttula.php*** for more details about N-Gage Arena and the N-Gage experience.

- monitoring their ranking on the score boards. When setting up a multiplayer session for the game, the user can choose whether to play with someone of similar rank or play freestyle, with whomever happens to be available to play

- using the chat and message boards

- sharing custom content they have created, such as photos and avatars

- checking their friends list to see who is online and ready to play a game

- updating their player profile.

N-Gage games rely on the server infrastructure shown in Figure 5.3 to provide scalable network application package (SNAP) game services, instant messaging and presence service (IMPS), and web services.

- SNAP Game services provide connected, multiplayer functionality to games with high performance, scalable event routing. For example, when one player takes a turn in a game, SNAP notifies the other Arena clients in the game, which in turn notify the game code and thus the player.

- IMPS Backbone Services support instant messaging (in-game or outside of games) and presence services. IMPS uses standard XMPP/Jabber protocols, extended in places to add custom Arena features.

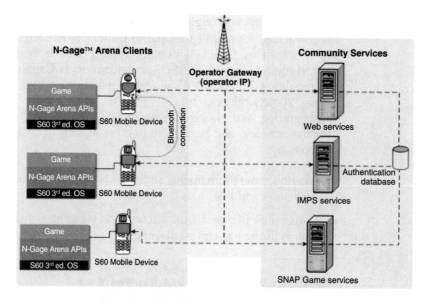

Figure 5.3 The N-Gage Arena system architecture

- Web services provide user account management. All web service data is centralized for use by game applications. The web services are the backbone of the Arena community features. Standard ports and protocols are used for flexibility and portability, for example, to enable multiplatform games to communicate.

If you've already taken a look at Chapter 9, you'll see a similar diagram in section 9.10, where Sam talks about the use of SNAP for multiplayer games. The Arena client API for N-Gage games is the C++ equivalent of the public SNAP SDK for Java ME developers. Arena and SNAP share the same backend framework, but the Java ME SNAP client uses an Apache-based proxy, while native clients contact the server backend directly.

A more significant difference is that the Java ME SNAP SDK is available for general use, and works on almost any mobile device supporting Java ME, not just Symbian OS smartphones delivered by Nokia. The APIs to access the Arena client are available only to N-Gage developers and are supported only on S60 3rd Edition smartphones.

The backend Arena services are hosted by Nokia on server clusters in several locations worldwide. It is expected that all N-Gage game titles will support a common minimum set of N-Gage Arena features, such as the in-game notifications discussed in Chapter 8. Some games will support real-time online multiplayer features, while others will use turn-based play or other innovative ways to create games that take advantage of the natural connectivity inherent in a mobile phone. A good example of a new type of mobile multiplayer game that exploits the N-Gage Arena APIs is **_Project White Rock_** which was briefly described in the section on multiplatform games in Chapter 1, section 1.5.3.

N-Gage games can also use custom game servers which run on the backend to provide additional, game-specific enhancements. Examples include:

- interaction with third-party software to share data between a game and, for example, a website

- community enhancements, such as the constant evolution of a virtual world required by a MMORPG

- faster data processing. The custom game server can perform processing tasks that would be inefficient or impractical if carried out on the phone handset.

A separate custom game server SDK is provided with the N-Gage SDK, for server creation. Custom servers can be written in C/C++, and Python bindings are also available for the game server client.

5.6 Other Online Multiplayer Solutions

Other solutions are available for mobile games to integrate support for online multiplayer functionality. For example, Terraplay provides the Global Gaming Network, which is an ASP service that enables game developers to tie their games to a network community, and to provide multiplayer gameplay. The Global Gaming Network supports mobile connected gameplay on all platforms, and also includes community and purchasing support, including billing.

SDKs are available for Symbian OS, BREW, and Java ME, and, at the time of writing, Terraplay reports that over 40 games, from 15 studios, have used their SDK.[7]

Exit Games Neutron 4.0 is another platform for development, deployment and hosting of mobile multiplayer games.[8] However, at the time of writing, the platforms supported included Java ME, BREW and Flash, but not native Symbian OS.

5.7 Further Reading

For additional information about communications programming on Symbian OS, we recommend Iain Campbell's *Communications Programming On Symbian OS (Second Edition)*, published by Symbian Press, 2007. The book includes a useful section on the development setup necessary to get you started debugging code that uses Bluetooth wireless technology using the Symbian OS emulator on Windows.

Forum Nokia has a selection of papers and examples about using Bluetooth wireless technology on S60 smartphones. These can be found at ***www.forum.nokia.com*** but have URLs far too long to type in manually. Please use the site search for the documents titled *S60 Platform: Bluetooth API Developer's Guide* and *S60 Platform: Bluetooth Point-to-Multipoint Example* or check the wiki site for this book, which will maintain a set of links to the relevant examples and papers available on the Internet.

Forum Nokia also has a number of legacy papers about mobile multiplayer game design, some of which are still very relevant at a high level. We recommend *Overview of Multiplayer Mobile Game Design* and *Introduction to Mobile Games Development*, both of which can be found by using the search option on the main page at ***www.forum.nokia.com***.

---

[7] Please see ***www.terraplay.com*** for more information.
[8] Further details can be found at ***www.exitgames.com***.

6

Exploiting the Phone Hardware

Paul Coulton and Fadi Chehimi
(www.mobileradicals.com)

6.1 Introduction

The standard game playing configuration has a game displayed on a screen and one or more players either standing or seated around it, interacting through a joystick and/or button interface. Although much argument reigns around who developed the first, of what were originally termed video games, this particular configuration became firmly established in the arcades of the late 1970s and early 1980s through classic game such as **Space Invaders** (1978 Taito), **Galaxian** (1979 Namco), **Pacman** (1979 Namco), and **Donkey Kong** (1981 Nintendo) to name but a few of our personal favorites. Whilst a mobile phone can be used for playing games in this manner, the very nature of the device, the wide user demographic, and its ever expanding feature set and capabilities, mean that developers can explore innovative interaction modalities and game playing experiences. It is to this premise that this chapter is dedicated.

In terms of game playing experience, Chapter 1 has already discussed the fact that the mobile phone gamer demographic is much more varied than that of console games, which means that developers should be encouraged to explore games outside the traditional genres. Mobile phones have literally become smartphones – sophisticated multimedia communications devices that we carry with us at all times and are integrated into very fabric of our daily activities. With these devices, we have the opportunity to create games that break out of a purely virtual experience and integrate readily into real world activities. Although for some of us, the mobile phone seems the ideal platform for these types of game experiences, it is not exclusively used, and more general definitions have arisen, such as mixed reality, augmented reality, alternate reality, and pervasive games. Before proceeding with this chapter, it is worthwhile exploring some of these areas.

6.1.1 Mixed Reality

Mixed reality games link the physical and digital worlds to create new experiences. The games often incorporate knowledge of their physical location and landscape, and then provide players with the ability to interact with both real and virtual objects within both the physical and digital worlds.[1] An example of a mobile phone mixed reality game is that of **PAC-LAN** shown in Figure 6.1 which is a version of the video game **Pacman** in which human players play the game on a maze based around the Alexandra Park accommodation complex at Lancaster University.[2] One player takes the role of the main **PAC-LAN** character and collects game pills (using a Nokia 5140 mobile phone equipped with a Nokia Xpress-on radio frequency identification (RFID)/near field communications (NFC) reader shell). The pills are in the form of yellow plastic discs fitted with stick-on RFID tags placed around the maze and are a direct physical manifestation of the virtual game pills on the mobile screen, placed at the real location corresponding to the virtual maze. Four other players take the role of the 'ghosts' who attempt to hunt down the **PAC-LAN** player. The game uses a Java ME application, running on the

Figure 6.1 *PAC-LAN* mixed reality mobile game

[1] Rashid O., Mullins I., Coulton P., and Edwards R. *Extending Cyberspace: Location Based Games Using Cellular Phones*, ACM Computers in Entertainment, Vol 4, Issue 1, January, 2006.

[2] Rashid O., Bamford W., Coulton P., Edwards R, and Scheible J. *PAC-LAN: Mixed reality gaming with RFID enabled mobile phones*, ACM Computers in Entertainment, Vol 4, Issue 4, October, 2006, pp 1–17.

mobile phone, connected to a central server using a GPRS connection. The server relays to the **PAC-LAN** character his/her current position along with the position of all ghosts, based on the pills collected.

6.1.2 Augmented Reality

Augmented reality games again mix a combination of real world and computer generated data although it is principally associated with the use of live video images which are processed and 'augmented' by the addition of computer generated graphics. In terms of mobile, we will deal with many aspects of this concept in section 6.2. But, as an example, consider Figure 6.2, which illustrates the mobile phone game **MobiLazer**.[3] The game uses specially designed colored-tags, which are worn by the players, and advanced color tracking software running on a camera phone, to create a novel first person shoot 'em up with innovative game interactions and play. The game is written in C++ for S60 smartphones and is controlled by a central server communicating via TCP/IP over GPRS.

Figure 6.2 **MobiLazer** augmented reality mobile game

6.1.3 Alternate Reality

Alternate reality games use interactive narratives with the real world as a platform to tell a story that may be affected by participants' ideas or actions and are often associated with theories of collective intelligence. The genre is typified by intense player involvement in real-time and evolves according to participants' responses. Characters are actively controlled by the game's designers, as opposed to being controlled by artificial intelligence (AI), as in a typical computer game.

Alternate reality games are increasing in popularity. They tend to be free to play, with costs absorbed either through supporting products

---

[3] Chehimi F., Coulton P., and Edwards R., *Augmented Reality 3D Interactive Advertisements on Smartphones*, The Sixth International Conference on Mobile Business, Ontario, Canada, 9th–11th July 2007.

(for example, **Perplex City** is funded by collectible puzzle cards, see **www.perplexcity.com** for more information) or through promotional relationships with existing products.

6.1.4   Pervasive Games

Pervasive games is another term often used to describe the types of games that are interwoven with our everyday lives through the items, devices, and people that surround us, and the places we inhabit. Much of the work in this area is associated with academic research initiatives, originally closely linked to ubiquitous and nomadic computing. Perhaps because of these links, they have only recently started to use mobile phones; examples being the SMS game **Day of the Figurine'**,[4] which could also be regarded as an alternate reality game and **Insectopia**,[5] which is based around Bluetooth proximity – a mechanism we shall discuss later in this chapter.

Whilst mixed reality games are an exciting prospect to create an engaging experience, we must also consider how the players will interact with objects and places, both real and virtual, using their mobile phone. In particular we must think beyond the standard phone keypad and four-way input controller and consider non-traditional input mechanisms such as a camera or 3D motion sensors, and how best to incorporate user context through location and/or proximity. This is the subject of the rest of this chapter.

6.2   Camera

Cameras are now a common feature of even the most basic mobile phone and provide developers with an interesting opportunity for their use within games. However, as yet, there have been relatively few examples of games that have done so, and they have used the camera in very different ways.

Other of the earliest of such games, unsurprisingly, came out of Japan around 2003, such as **Photo Battler** from NEC. This game allowed players to turn photos into character cards that were assigned various attributes, enabling them to compete against each other. At around the same time

---

[4] Flintham M., Smith K., Benford S., Capra M., Green J., Greenhalgh C., Wright M., Adams M., Tandavanitj N., Row Farr J., and Lindt I., *Day of the Figurines: A Slow Narrative-Driven Game for Mobile Phones Using Text Messaging*, Proceedings of 5th International Workshop on Pervasive Games, Salzburg, Austria, June 11–11, 2007.

[5] Peitz J., Saarenpää H., and Björk S. *Insectopia: Exploring pervasive games through technology already pervasively available*. In Proceedings of the International Conference on Advances in Computer Entertainment Technology, Salzburg, Austria, June 13–15, 2007.

Shakariki Petto appeared from Panasonic, which took the form of a virtual pet that a player fed by taking pictures of colors that represented food, for instance the color red represented apples. More recent games have also explored using the pictures themselves to create mixed-reality games. The ***Manhattan Story Mash-Up***[6] used players on the streets of Manhattan who were given words defined by other players, who were online, and had to take a picture to represent the word. Other players then voted on the most applicable picture, and the player who took that picture was awarded points. A similar concept was explored in ***My Photos are My Bullets***[7] although in this case, the object of the game was to take a picture of a prescribed opponent and then let an independent 'judge' decide the number of points awarded, based on the quality of the image.

Other games have evolved to use the camera to detect movements of the phone and transfer them to movements within the game. Probably the best known are from game developer Ojom (***www.ojom.com***) with its games ***Attack of the Killer Virus*** and ***Mosquitos*** for Nokia's S60 smartphones. In both games, the enemy characters that the player must 'shoot' are superimposed on top of a live video stream from the mobile phone's camera. The player moves around this mixed reality space by moving his phone and firing using the centre key of the joy pad. Although this technique sounds complex, it is a fairly straightforward piece of signal processing, whereby captured images are generally grid sampled, to reduce the complexity, and then some form of block matching algorithm is used on successive images to estimate direction of motion. However, it should be noted that the granularity of control is large, the technique is affected by camera quality and lighting levels, and it is extremely power hungry.

Some games have evolved to use visual codes to either detect movement, such as an augmented reality version of table tennis,[8] which is based on an implementation of the Augmented Reality Toolkit for Symbian OS. The system interacts with code markers fixed on a physical object to calculate the rotational vectors it applies to a pre-rendered image. Figure 6.3 shows an example of such a system, but for adverts not games, which improves upon previous versions in that the 3D object is stored on the specially designed tag and rendered on the fly. However,

[6] Tuulos V., Scheible J., and Nyhom H., *Combining Web, Mobile Phones and Public Displays in Large-Scale: Manhattan Story Mashup*, Fifth International Conference on Pervasive Computing, Toronto, Ontario, Canada,13–16 May 2007.

[7] Suomela R., and Koivisto A., *My photos are my bullets – Using camera as the primary means of player-to-player interaction in a mobile multiplayer game*, International Conference of Entertainment Computing 2006, 20–22, September 2006, Cambridge, UK, pp 250–261.

[8] Henrysson A., Billinghurst, M., and Ollila M., *Face to Face Collaborative AR on Mobile Phones*. In Proceedings of the International Symposium on Mixed and Augmented Reality (ISMAR 2005), October 5th–8th, 2005, Vienna, Austria.

Figure 6.3 Using cameras to capture motion

all these systems suffer similar problems to those experienced by optical flow techniques.

Other games have used various forms of a two-dimensional bar code; the most famous example being **ConQwest** by Area Code, ***www. playareacode.com***, which used Semacodes. **ConQwest** was a team-based treasure hunt game using implied positioning from Semacode stickers and was sponsored by Qwest Wireless in the USA to promote its camera phones. Each sticker was given a relative value, and the players collected the stickers by taking pictures with their camera phones. The first team to collect $5000 worth of Semacodes won. The game was played by teams, generally drawn from local high schools in various parts of the USA.[1] More information about the use of bar codes in mobile games can be found in section 6.3.6.

6.2.1   Using the Camera on Symbian OS

When writing native C++ games for Symbian OS platforms, use of the camera requires the UserEnvironment platform security capability. The CCamera API is used, and one of the two available callback interfaces, MCameraObserver or MCameraObserver2, should be implemented. The primary difference between the interfaces is in the way the camera is managed when several applications use it simultaneously. MCameraObserver2 introduces a priority level registered for each application which determines the ownership of the camera hardware at an instant of time. However, we will use MCameraObserver in the example given below.

Symbian OS provides an abstract onboard camera API, called ECam, which is implemented by handset manufacturers according to the capabilities of each phone. ECam provides a hardware agnostic interface for applications to communicate with, and control, any onboard camera hardware, via the implementation of the CCamera API provided by the manufacturer. Recent releases provide enhanced control of different camera settings, given they are supported by the hardware of the camera, such as introducing camera stabilization functions, ISO rate controller, and various focus and aperture setting options. These are implemented in class CCameraAdvancedSettings which is defined inside class CCamera. As we shall only deal with the basic components of the camera API of Symbian OS v9.1 in this chapter, we refer readers to the relevant platform SDKs for a detailed discussion on using the advanced features.

Below, we provide an example using MCameraObserver which is called *ColourTracker*. It enables a Nokia S60 3rd Edition smartphone, to identify a specific color in a scene and then track it, as shown in Figure 6.4. While this example merely provides the tracking functionality, one could easily imagine it being used for identifying targets in a mixed reality shooting game where players use their camera phones to detect targets by their colors.

Figure 6.4 Screen shots of *ColourTracker*. The left image shows a red London Route Master bus before processing. The right image is after processing, where the red color has been identified, tracked and replaced with white

Camera API Overview	
Library to link against	`ecam.lib`
Header to include	`ecam.h`
Required platform security capabilities	`LocalServices, UserEnviroment`
Classes to implement	`MCameraObserver`
Classes used	`CCamera`

`CCamera` provides several asynchronous functions that enable applications to control the cameras on mobile phones. These functions are asynchronous to prevent the application thread from being blocked while waiting for code to finish executing, such as when capturing a camera frame.

The initial step in using the camera in Symbian OS is to construct it, which is done by calling the static factory function `CCamera::NewL()`. This may result in a panic with one of the following error codes generated:

- `KErrNotSupported`, if no camera is installed on the phone or if an incorrect camera index is passed as a parameter (for example passing 1, which indicates the front camera, when the phone has only a back camera, which is index 0)

- `KErrPermissionDenied`, if the `UserEnvironment` capability is not listed in the application's MMP file

- `KErrNoMemory`, if insufficient memory is available for allocating the `CCamera` object.

Not all Symbian OS phones come with cameras built-in, so it is essential to handle the case where camera-based applications, such as **ColourTracker**, are installed on phones without a camera. For this reason it is good practice to always check the availability of the camera by calling `CCamera::CamerasAvailable()` before implementing any camera-dependent code. The function returns the number of cameras installed on the phone and returns zero if no camera is present.

Assuming that there is at least one camera on the smartphone, the next step is to reserve the camera to grant the client application a handle to it. To perform this task, the asynchronous `CCamera::Reserve()` function must be called on the `CCamera` instance. Once the camera is reserved, it has to be switched on by calling another asynchronous function `CCamera::PowerOn()`. The camera then will be ready to receive data from the camera viewfinder in bitmap buffers via either of the activation functions, `CCamera::StartViewFinderBitmapsL()` or `CCamera::StartViewFinderDirectL()`.

Of the two functions, `CCamera::StartViewFinderDirectL()` is the faster, as it transfers the viewfinder buffer from the camera to the screen directly using direct screen access (DSA), while `CCamera::StartViewFinderBitmapsL()` does the drawing through the window server (WSERV). In the example below we will be using `CCamera::StartViewFinderBitmapsL()`.

The client application needs to inherit from the interface class `MCameraObserver` in order to actively interact with the camera API. This mixin class provides five pure virtual functions that must be implemented:

- `void ReserveComplete()`: handles the completion result from camera reservation

- `void PowerOnComplete()`: indicates whether camera power-on completed successfully

- `void ViewFinderFrameReady()`: receives the data from the camera viewfinder bitmap

- `void ImageReady()`: handles the completion event raised by image capture

- `void FrameBufferReady()`: handles the completion event raised by video capture.

Having discussed the basic set-up we can now proceed with our specific example. In its header file, `ColorTrackerView.h`, the following can be seen:

- inclusion of `ecam.h` for the camera APIs and `FBS.h` for bitmap manipulation (also requires linking against `fbscli.lib` in the project's MMP file)

- inheritance from `MCameraObserver`

- declaration of the tracking function `TrackColor()`

- declaration of the `iCamera` and `iViewFinderBitmap` member variables (the latter is a bitmap holding individual frames from the camera) and a Boolean member variable (`iAllowDraw`) to indicate the availability of a camera frame from the viewfinder.

```
#include <ecam.h>          // Camera API
#include <FBS.h>           // Fonts and Bitmap API

class CColorTrackerAppView : public CCoeControl, MCameraObserver
```

```
    {
public:
    ... // Construction, destruction omitted for clarity
private:
    // Inherited from MCameraObserver
    void ReserveComplete(TInt);
    void PowerOnComplete(TInt);
    void ViewFinderFrameReady(CFbsBitmap&);
    void ImageReady(CFbsBitmap*, HBufC8*, TInt);
    void FrameBufferReady(MFrameBuffer*, TInt);

    // The function to track colors in our example
    // @arg1: The color to track
    // @arg2: Reference to the application's graphics context
    void TrackColor(TRgb&, CWindowGc&);
private:
    CCamera* iCamera;
    CFbsBitmap* iViewFinderBitmap;
    TBool iAllowDraw;
    };
```

On instantiation, as part of the second phase construction of a `CColorTrackerAppView` object, the `ConstructL()` method constructs the `iCamera` object and reserves it, and constructs the viewfinder bitmap `iViewFinderBitmap`.

```
void CColorTrackerAppView::ConstructL(const TRect& aRect)
    {
    ...// Omitted for clarity
    // Instantiate iCamera:
    // arg1: reference to the inherited MCameraObserver interface
    // arg2: the index of the device camera to use - 0 for the
    // back camera.
    iCamera = CCamera::NewL(*this, 0);

    // Check if there is a camera
    if( iCamera->CamerasAvailable() > 0 )
        {
        iCamera->Reserve();
        // Construct the bitmap to hold the frames captured by the camera
        iViewFinder = new(ELeave) CFbsBitmap();
        }
    else
        {
        User::Leave(KErrNotFound);
        }
    }
```

As we previously discussed `CCamera::Reserve()`, `CCamera::PowerOn()`, and `CCamera::StartViewFinderBitmapsL()` are all asynchronous functions, and each must only be called when the preceding function in the sequence has completed, because dependencies exist. For example, `StartViewFinderBitmapsL()` is dependent on the completion of `PowerOn()`, which is itself dependent on the completion of `Reserve()`. If `StartViewFinderBitmapsL()` is called before the `PowerOn()` method completes, it will panic the application.

To manage this sequence of calls, we implement the callback functions of the `iMCameraObserver` to provide the framework to monitor the completion of each camera operation. Therefore, we add the following function implementations to `ColorTrackerView.cpp`.

```cpp
void CColorTrackerAppView::ReserveComplete(TInt aErr)
  {// Check the camera is reserved successfully
  // Error handling is omitted for clarify
  if( aErr==KErrNone)// Turn the camera on
    iCamera->PowerOn(); // Asynchronous completion
  }

void CColorTrackerAppView::PowerOnComplete(TInt aErr)
  {
  if (aErr==KErrNone)
    {// Set the size of the rectangle on screen to use
    // for the view finder
    TSize size(240,180);
    // Start capturing frames -  asynchronous completion
    TRAPD( result, iCamera->StartViewFinderBitmapsL(size) );
    if( result != KErrNone)
      {// Error handling omitted for clarity
      }
    }
  }

void CColorTrackerAppView::ViewFinderFrameReady(CFbsBitmap& aFrame)
  {
  // There is one frame ready to display,
  iAllowDraw = ETrue;
  // duplicate it,
  iViewFinder->Duplicate(aFrame.Handle());
  // and render it on screen.
  DrawNow();
  }

void CColorTrackerAppView::ImageReady(CFbsBitmap*, HBufC8*, TInt)
  {
// No need for implementation.
  }

void CColorTrackerAppView::FrameBufferReady(MFrameBuffer*, TInt)
  {
// No need for implementation.
  }

void CColorTrackerAppView::Draw(const TRect& aRect) const
  {
  ...// Omitted for clarity
  // Allow bitmap drawing and manipulation
  if(iAllowDraw)
    {
    // The color to track (red)
    TRgb colorToTrack(255,0,0);
    // Draw the frame on screen
```

```
gc.BitBlt(TPoint(0,0), iViewFinder);
// Track the red color
TrackColor(colorToTrack, gc);
}
}
```

When the camera reservation is finished, an event is raised that results in a call to `MCameraObserver::ReserveComplete()` to notify it of completion. On checking that there is no error passed as a parameter to the callback function, to confirm that reserving the camera was successful, `PowerOn()` is called from within the handler function, which starts powering up the camera asynchronously. If this completes successfully we call `StartViewFinderBitmapsL()` from `PowerOnComplete()`, which starts loading the frames from the camera viewfinder.

When the first bitmap frame is ready, the `MCameraObserver::ViewFinderFrameReady()` observer function is called by the framework, passing as a parameter that frame (`aFrame`), allowing it to be copied, or duplicated, into `iViewFinderBitmap` for later manipulation. The method sets `iAllowDraw` to `ETrue` to indicate that it is possible now to start drawing bitmap frames to the screen as there is at least one that has been successfully loaded. `DrawNow()` in `ViewFinderFrameReady()` calls `Draw()` to perform the on-screen drawing. You should not call `Draw()` directly as there are several steps WSERV takes before and after a call to `Draw()` which `DrawNow()` deals with internally.

The previous code illustrates the use of the camera on any Symbian OS smartphone. To complete this example, the following function performs the actual tracking of a color, red in our example. The function is called from within `Draw()`, which will draw white squares on the screen in place of the red pixels. Note that this is a very simple color tracking algorithm, and we acknowledge that there are more sophisticated and accurate solutions available.

`TrackColor()` identifies a threshold for the color to track in a spherical shape surrounding it. The function calculates the distance between the color to track, represented as the centre of the sphere, and the color of an arbitrary pixel on screen in the RGB ($0 \leq R$, G and $B \leq 255$) space, as shown in Figure 6.5. Point C represents the centre of the color threshold. R is the radius of the sphere and r is the distance between B (which falls within the boundary) and C. Apparently point A is outside the sphere boundary making L much beyond the length of R, and thus A is not considered for detection.

If the distance between the color of an arbitrary screen pixel and the color to track (in the RGB space) is less than the predefined radius of the surrounding sphere, 125 in our example, then the two colors match, and that pixel is replaced with white. Otherwise, the color is not recognized

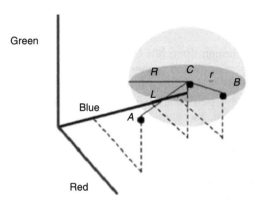

Figure 6.5 Spherical boundary color detection

and the functions continue checking other pixels. The following code implements this detection method.

```
void CColorTrackerAppView::TrackColor(TRgb& aColor, CWindowGc& aGc)
  {
  // Set the color of the dots to draw
  aGc.SetBrushColor( KRgbWhite );

  TUint r, g, b, distance;
  TRgb pixelColor;

  // Traversing the 240x180 viewfinder bitmap captured
  for(TUint x=0; x<240; x+=2)
    {
    for(TUint y=0; y<180; y+=2)
      {
      iViewFinder->GetPixel(pixelColor,TPoint(x,y));
      r = pixelColor.Red();
      g = pixelColor.Green();
      b = pixelColor.Blue();

      // Calculating the threshold as a sphere radius. This uses
      // Pythagoras theorem in calculating lengths.
      distance = (r-aColor.Red())*(r-aColor.Red()) +
        (g-aColor.Green())*(g-aColor.Green()) +
        (b-aColor.Blue())*(b-aColor.Blue());

      // If target color is within 125 radius
      if(distance <= 15625)   // 125^2
        {
        aGc.DrawRect(TRect(TPoint(x,y), TPoint(x+2, y+2)));
        }
      }
    }
  }
```

This completes our admittedly simple example that illustrates a little of the potential of the camera API provided by Symbian OS for mobile game developers.

6.3   Location

Although there has been a lot of publicity in Europe and the US about location-based games, they are still considered by many as a fringe genre. There are a few reasons for this: first, it is an immature industry that is particularly susceptible to the fragmentation of mobile standards; second, the games often require a critical mass of players to become successful, which is difficult to obtain within the limited distribution options available. Whilst we won't be addressing questions relating to generating revenue from such games in this chapter, we can point out that infrastructure and handset support for location-based games is growing all the time, and that some games described have generated high levels of public interest.

Providing location information is far from straightforward, often requiring significant changes in the software and hardware of the mobile system and/or the handset to ensure that the information is relevant to the user's actual position. Location information is essentially a two-stage process, in that we have to provide both the geographical position of the mobile user and the information on a particular product or service related to that location. In the following section, we will highlight the location positioning technologies available to mobile games developers and highlight some of the location-based games which have utilized some of these methods.

6.3.1   Obtaining Location

Often those new to the subject consider finding the location of a mobile phone user as a new problem, whereas, in fact, all mobile phone systems effectively track a user's whereabouts at the cellular level. Each cell site has a unique id that enables the particular mobile phone network system to locate a mobile phone subscriber, so that it can route calls to the correct cell. However, to improve on this relatively crude accuracy, other techniques treat location-finding as a relative exercise; in other words, the location of the mobile phone user must be estimated against some known framework. This framework could be the locations of the base stations of a mobile phone network or the satellites of the global positioning system (GPS). Each system will provide different levels of performance and capabilities, and in the following paragraphs we discuss the relative merits of the various methods together with highlighting mobile games developed around these techniques.

6.3.2   Cell ID

From the previous discussion, the most obvious location method is that of the GSM cell ID, as it requires no modifications to either the phone handset or the network infrastructure. While the cell ID has a typical location acquisition time of around three seconds, its accuracy depends upon the size of the cell, which varies depending upon the capacity requirements of a particular geographical area (for 3G systems this could range between a cell radius of 100m to 10 km).

Whilst there are a numbers of location-based games using cell ID, in this discussion we will consider probably the most well known, which is **Botfighters** (**www.botfighters.com**). The game was launched in 2000 by a Swedish company called It's Alive as a search and destroy combat game. It is fairly representative, in terms of technology and gameplay, of a number of other games in this genre such as **UnderCover** and **Gunslingers**. In the first version of **Botfighters**, players sent an SMS to find the location of another player and if that player was in the vicinity, they could shoot the other player's robot by sending another SMS message. While gaining some initial recognition, **Botfighters'** use of SMS and its crude location accuracy meant that it suffered from delay and minimal levels of social interaction, as players were physically distributed over large areas. **Botfighters 2** attempted to remedy this situation by using a Java ME client and GPRS connectivity to upgrade gameplay to incorporate collaborative game playing, character upgrading, weapons and bonus material trading, as well as AI-controlled opponents. Despite its aspiration to become a massively multiplayer online role playing game (MMORPG), a greater degree of location accuracy and social interaction would undoubtedly be required to attain this goal.[1]

6.3.3   Time of Arrival

To estimate location, time of arrival (TOA) systems measure the differences among the multiple signals arriving at a device. In order to enable a two-dimensional calculation (i.e., latitude and longitude), TOA measurements must be made with respect to signals related to at least three geographically distinct framework elements. These measurements allow us to infer either the absolute or the differential distance of signal propagation between the framework elements and the device. A system that is based on differential distance estimation is generally referred to as time difference of arrival or TDOA. TOA requires a common timing base, which is fine for synchronous mobile systems like cdmaOne, but for non-synchronous systems such as GSM, special timing units must be added to the infrastructure. The accuracy of such systems can be between 50 and 250 meters, and is extremely susceptible to multipath accuracy problems. Therefore, we have yet to see such systems implemented for the general consumer and, as such, TOA has been unused for games.

6.3.4   Global Positioning System (GPS)

GPS was originally developed by the US military for defense purposes; it is based on a system of 24 satellites that orbit the earth. A GPS receiver can triangulate its position using TOA methods, as long as at least three satellites are visible. Typically, accuracy is around 2 to 10 meters. The cost of GPS hardware continues to fall, and GPS remains a free service to consumers. Integrating GPS hardware technology into mobile phones is seen as an increasingly viable solution, as can be seen with devices such as the Nokia N95. However, GPS has a number of major restrictions, such as limited coverage in urban environments and inside buildings due to the requirement that mobile users be in 'view' of the satellites, and a slow location acquisition time (in the region of 10 to 60 seconds).

There have been a number of mobile GPS games, the most famous being **Mogi Mogi** which was a treasure hunt game developed by Newt Games, for the Japanese mobile operator KDDI to publicize the launch of their eight new GPS-enabled handsets in 2003. The game was based around a Java ME application that allowed players to accumulate as many points as possible by collecting virtual treasure items spread randomly over a virtual map of Tokyo. Items could be collected once the user reached within 400 meters of the virtual item and players could meet other players to trade items. The location tracking was provided by either GPS or cell ID, and the developers recommended that GPS should only be used when looking for a particular item or players in a small area. Cell ID is used to get a general bearing to reduce heavy battery consumption. **Mogi Mogi** is by no means unique in this area and other pioneering GPS mobile games worth mentioning are **Raygun** by Glofun and **Undercover 2** by YDreams

6.3.5 Assisted GPS (A-GPS)

Assisted GPS systems overcome some of the limitations of GPS position acquisition by using the fixed mobile network infrastructure to reduce acquisition time to less than 5 seconds, and may provide indoor accuracy to within 50 meters. However, as with infrastructure TOA schemes, multi-path interference and the lack of a line of sight measurement can degrade performance, so A-GPS can be less accurate than GPS under certain conditions. For network operators, the main benefit from integration of A-GPS is the ability to charge users for positioning information, because it uses the network infrastructure.

The use of A-GPS systems in games has been seen predominantly in the US. An example is **Swordfish**, which is a location-based fishing game developed by Blisterent, where players use their mobile phones and A-GPS to find and catch virtual fish. A player's general movement in his or her local area will bring the player into the vicinity of the fish, and the player can then move to a particular location to try and catch the fish. The positions are all relative, in that they do not have to be in a particular city or area to play the game. **Torpedo Bay** is by the same developers and utilizes the same methodology as *Swordfish*, but it is essentially a multiplayer game, in a similar vein as the old battleships game but using street maps.

6.3.6 Implied Location

Implied location solutions are systems where the mobile phone user can interact with objects or systems that have a known location relative to the mobile network infrastructure. Thus a user's particular location can be implied by his or her interaction with the objects or systems. At the system level, this could be the interaction with WLAN cells or Bluetooth piconets. WLAN cells have a relatively small coverage area, up to 100 meters, and as some mobile phones already include WiFi functionality, it could be used to provide a general location area; alternatively, there have been proposals to use signal strength and the interaction between different access points to obtain greater accuracy. However, the use of signal strength has a number of limitations which make its use outside a very defined environment, such as a research laboratory, somewhat impractical.[1] Bluetooth technology offers a similar approach for general location identification, although within a smaller area than WiFi, at around 10 meters. With its high penetration of the consumer market, greater robustness against interference, and low power requirements, Bluetooth is arguably a much more practical solution for mobile phones and has indeed started to be used for localized advertising.

In terms of interacting with objects to determine a user's location, there are two main possibilities: two-dimensional bar codes, and RFID/NFC tags. Two-dimensional bar codes come in a variety of forms: Quick Response (QR) codes (an example of which is shown in Figure 6.6),

Figure 6.6 An example QR bar code

Semacodes, and ColorCodes, to name but a few of the variants. All of these variants can contain an Internet address, which, when scanned, prompts the phone to load the relevant page. In Japan, QR codes have become commonplace on business cards to allow people to easily upload contact details onto phones, or in some cases, allow blogging of pictures associated with physical locations, for example, **www.tokyo-picturesque.com**. All these codes require mobile phones with inbuilt cameras to take an image of the code, which is either decoded on the phone with specialist software, or transferred to an Internet service for decoding. In terms of games, the only example of note is the **ConQwest** game discussed in section 6.2.

RFID/NFC tags are devices that can transmit a radio frequency signal that contains information about the object to a suitable reader. ABI Research has predicted that RFID-enabled phones will occupy 50% of the global market by 2009. That fact, coupled with its very low power operation, make it a suitable candidate for games, as previously highlighted by **PAC-LAN**. The results of the research from **PAC-LAN** and two other games, **Mobspray** and **MobHunt**,[9] shown in Figure 6.7, indicate that RFID/NFC tags are superior to visual tags because they have faster read times, as the tags can be accessed at rates in excess of 100 kbits/s, whereas the two-dimensional bar codes require image capture and processing which typically takes few seconds. In addition, RFID tags can be written to, as well as read from, and have a simpler reading method, as the phone and the tag have merely to 'touch' or be placed in close proximity (less than 3 cm), whereas the bar codes require

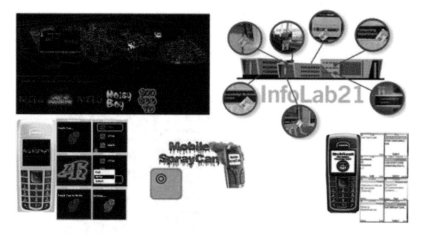

Figure 6.7 *Mobspray* and *MobHunt* games

[9] Coulton P., Rashid O., and Bamford W., *Experiencing "Touch" in Mobile Mixed Reality Games*, Proceedings of The Fourth Annual International Conference in Computer Game Design and Technology, Liverpool, 15th – 16th November 2006, pp 68–75.

the user to take a picture. RFID tags are also more robust, as errors are more likely to occur when scanning a bar code due to irregular camera orientation.

6.3.7 Using Location on Symbian OS

To allow developers to program location-aware applications more easily, Symbian introduced the location-based services (LBS) API, or location acquisition (LA) API as it is sometimes called. LBS was initially a Nokia component introduced into S60 2nd Edition, Feature Pack 2, but it later found its way to the core of the Symbian OS platform in Symbian OS v9.1. The API is comprehensive and broad enough to cover both network based positioning and GPS methods, to allow flexibility in applications. In particular it provides support for:

- accurate coordinates identification with error estimates

- bearing to, distance, altitude, latitude, and longitude determination

- speed of the device and its direction of movement measurements (with accuracy estimates)

- GPS network services and satellite information

- Quality of Position (QoP) (dilution of precision) information estimations.

Location-Based Services API Overview	
Library to link against	`lbs.lib`
Header to include	`lbs.h`
Required platform security capabilities	`Location`
Classes used	`RPositionerServer` `RPositioner` `TPositionInfoBase` `TPositionUpdateOptions`

In this section, we demonstrate how to use the LBS API with GPS and provide a simple application that prints the latitude and longitude of the smartphone onto the screen. The example is the same for mobile phones with either internal or external GPS as the API automatically searches for an internal GPS receiver once started and if none is found it then searches for an external receiver by initiating a Bluetooth discovery session. The user is then presented with a list of all found Bluetooth devices, including

Figure 6.8   Screenshot of **FindMe** on an S60 3rd Edition smartphone

Bluetooth GPS receivers, to choose from, and once chosen, it will be stored by the API for future quick reference.

The example is called **FindMe**, and a screenshot is shown in Figure 6.8. Note that the example requires the `Location` capability.

The LBS API consists of almost 30 classes which include base classes, server and position classes, location information classes, and many others. For the sake of simplicity, in this example, we are going to use only the four most basic classes needed to acquire the phone location, which are:

- `RPositionerServer` to make the primary connection to the location server

- `RPositioner` to create a subsession to the server in order to obtain current position information

- `TPositionInfoBase`, to hold the position information

- `TPositionUpdateOptions` to define the interval length for updates.

Since most LBS classes are R classes they are not constructed with the usual call to `NewL()`. They are simply defined as member variables of a class and connected, manipulated, and finally disconnected in its implementation.

As with any Symbian OS client-server API, a client application has to set up a connection to the LBS server by calling `RPosition-Server::Connect()` to enable it to retrieve positional information. Then the client must initiate a subsession to the connected server instance by using one of the `OpenL()` method overloads of the `RPositioner` class. Once the connection is established, a requestor for the location service is identified in order to use it.

A requestor guarantee is provided by the positioning framework to help prevent the misuse of positional information that is relevant to users' privacy and security. A requestor can be either a service or a contact. A service requestor represents any terminal or network application or service, and a contact requestor represents an individual person requesting the location. Since *FindMe* is an application (or service) on the phone, we utilize service requestors.

Once the requestor is set up, the final stage in implementing LBS for an application is to determine how often to capture the positional data updates, how long we should wait to receive the data in event of delay, and how long we should keep the cached data. Once this is determined, we direct the server to start updating our application with position data with a call to the asynchronous function `RPositioner::NotifyPositionUpdate()`. This takes as parameters a reference to an object of type `TPositionInfoBase` which holds the location information retrieved from the server, and a reference to an active object's `iStatus` member. Before terminating the application, we have to close both the subsession and the server connection with calls to `RPositioner::Close()` and `RPosition-Server::Close()` respectively.

Having described the basics for an LBS application, we can now turn our attention to the specifics of our example. LBS requires the implementation of an active object to manipulate the asynchronous calls for location information acquisition. For this purpose, we have implemented the active object class `CFindMeActive`, which is derived from `CActive`, implementing its two pure virtual functions `RunL()` and `DoCancel()`. The class encapsulates all necessary information and constructions for operating the location service. The class is declared as follows in `FindMeActive.h`:

```
#include <lbs.h>//Location-based services API

class CFindMeActive : public CActive
    {
    // Public methods for construction and destruction
    // are omitted for clarity
private: // Inherited from CActive
    // Handle request completion
    void RunL();
    // Clean up a request
    void DoCancel();
private:
    // Member variables
    RPositionServer iPosServer;
    // A pointer to our application's view class
    CFindMeAppView* iView;
    // Position information holder
```

```
  RPositioner iPositioner;
public:
  // Will be used from the view class
  TPositionInfo iPositionInfo;
  };
```

The second-phase `ConstructL()` method performs the main actions to connect to the server as previously discussed. The important thing to note is the use of function `RPositioner::SetRequestor()`, which is responsible for assuring the security of the personal location information of the user and takes three parameters:

- the type of the requestor `ERequestorService` or `ERequestor-Contact`

- the format of the identification of the requestor (`EFormatApplication`, `EFormatTelephone`, `EFormatUrl` or `EFormatMail`)

- the value of the requestor identification which can be the application name, telephone number, URL, or email address, depending on what was passed as the second parameter.

```
void CFindMeActive::ConstructL(CFindMeAppView* aView)
  {
  // Set the pointer to our application view
  iView = aView;
  // Connect to the position server
  User::LeaveIfError(iPosServer.Connect());
  // Open the subsession to the server
  User::LeaveIfError(iPositioner.Open(iPosServer));
  // Set the requestor
  User::LeaveIfError(iPositioner.SetRequestor(
                     CRequestor::ERequestorService ,    // Service type
                     CRequestor::EFormatApplication ,   // App format
                     KRequestor));                      // Our application name
  // Set update interval to receive position information
  // every 1 second.
  TPositionUpdateOptions udOpt;
  udOpt.SetUpdateInterval(TTimeIntervalMicroSeconds(KUpdateInterval));
  // The position server will terminate the request by the
  // requestor in case it could not retrieve position info
  // in 1 minute.
  udOpt.SetUpdateTimeOut(TTimeIntervalMicroSeconds(KUpdateTimeOut));
  // 5 seconds is the maximum age for a cached info.
  udOpt.SetMaxUpdateAge(TTimeIntervalMicroSeconds(KCacheAge));
  // Do not allow location server to send partial position data.
  // We are interested in complete data.
  udOpt.SetAcceptPartialUpdates(EFalse);
  // Set the update options.
  User::LeaveIfError(iPositioner.SetUpdateOptions(udOpt));
  // Set the active object to start getting asynchronous
  // position information.
```

```
iPositioner.NotifyPositionUpdate(iPositionInfo,iStatus);
SetActive();
}
```

When `ConstructL()` is executed, the active object submits an asynchronous request to receive positional information from the server. When positional data is available, `RunL()` is called by the active scheduler (which schedules all current active objects) to handle this obtained data. The function checks whether the position was updated successfully, before requesting the view to update the screen with this newly available positional data. It also issues a new request to get updated positional data. The application continues retrieving location data in this manner until it shuts down:

```
void CFindMeActive::RunL()
  {
  // If successfully received position info
  if( iStatus.Int() == KErrNone ||
      iStatus.Int() == KPositionPartialUpdate )
    {// Update the view
    iView->PrintPos();
    // Request the next position
    iPositioner.NotifyPositionUpdate( iPositionInfo, iStatus );
    SetActive();
    }
  ...// Error handling code omitted for clarity
  }
```

Active objects implement the `DoCancel()` function to cancel any outstanding asynchronous requests issued by the active object, and perform any necessary cleaning up. In *FindMe*, we cancel the notification request in `DoCancel()` and close the allocated sessions in the destructor.

```
void CFindMeActive::DoCancel()
  {
  iPositioner.CancelRequest(EPositionerNotifyPositionUpdate);
  }

CFindMeActive::~CFindMeActive()
  {// Cancel any outstanding requests - calls DoCancel()
  Cancel();
  iPositioner.Close();
  iPosServer.Close();
  }
```

The remaining action is to update the screen with the position data received from the server. This is done via `Draw()` and `GetDegreesString()` in class `CFindMeAppView`. `CFindMeAppView` uses the `iPositionInfo` member of class `CFindMeActive` to get the position data. The latitude and longitude fields are extracted from it next and passed to `GetDegreesString()` to format them appropriately for on-screen display. `GetDegreesString()` has been reused from the **LocationRefAppForS60** sample application in Nokia's S60 3rd Edition FP1 SDK.

```
void CFindMeAppView::Draw( const TRect& /*aRect*/ ) const
  {
  // If there is new location information found.
  if( iPosDataFound )
    {// Get the position information and store it in a
    // TPosition object.
    TPosition pos;
    iMyPos->iPositionInfo.GetPosition(pos);
    TBuf<20> lat;
    GetDegreesString(pos.Latitude(),lat);
    TBuf<20> lon;
    GetDegreesString(pos.Longitude(),lon);
    ...// Write lat and long to the screen.
    }
  }
```

6.4 3D Motion Sensors

The Nintendo Wii has achieved phenomenal sales since its launch, largely by capturing the imagination of players not traditionally considered part of the current game market. From the perspective of both the game developer and game player, the most innovative feature of the Wii is the 'Wiimote.' The Wiimote itself was one of the primary design aspects of the Wii, and, in an interview by Kenji Hall for *Business Week* in November 2006,[10] Shigeru Miyamoto describes part of Nintendo's rationale:

"The classic controller was something we had become fond of and gamers had become comfortable with. It had many important elements. But it also had come to dictate a lot of what went into games – the way graphics were made, the way battles were fought in role-playing games, the arc of in-game stories. They were all being made to fit one standard. Creativity was being stifled, and the range of games was narrowing...

"There are examples of controllers that were made for specific games, such as Konami's Dance Revolution. And for a long time, we thought that changing the interface would broaden game design and loosen creative constraints on programmers. We found that to be true when we released the DS. Around that time, we were also agreeing that we would start from the drawing board with something entirely unlike anything we had made before."

Their solution was the Wiimote, a wireless controller which is able to sense both rotational orientation and translational acceleration along

[10] *The Big Ideas Behind Nintendo's Wii*, BusinessWeek, November 2006 **www.businessweek.com/technology/content/nov2006/tc20061116_750580.htm**.

three dimensional axes. It achieves this through the use of inbuilt accelerometers, together with a light sensor. This light sensor is used in conjunction with an array of light emitting diodes centrally positioned above or below the console's display (see *forums.nintendo.com* for more information). The combination of Wiimote's accelerometer and light sensor data allows for six degrees of freedom. The Wiimote can be augmented with additional features, one of which is the amusingly titled 'Nunchuk,' which features an accelerometer and a traditional analogue joystick with two trigger buttons. The overall result is a game interface capable of supporting a huge array of input possibilities that will enable new and exciting video game experiences.

Although the first 3D accelerometers have only just been integrated into mobile phones and these phones have yet to become widespread they offer an exciting new direction for game design on mobile phones. The Nokia 5500 is one of the first such equipped phones having been targeted at sports users, using the built-in motion sensors as both a pedometer and speed/distance tracker for various exercising purposes. Other phones with motion sensors include Samsung's SCH-S310 and NTT's N702.

As we went to press, Nokia announced that, in later N95 firmware releases, developers can access raw accelerometer data from a device's inbuilt tri-axis accelerometer, using the official sensor plug-in downloadable from Forum Nokia (search for 'Sensor Plug-in for S60 3rd Edition SDK' at the main page of *www.forum.nokia.com*). Nokia Research Center has also developed an accelerometer software plug-in specifically for the Nokia N95. The plug-in is a C++ software library, which can be used with public S60 platform SDKs, although it is not an official release from Forum Nokia and is meant for research and development purposes only. For more information, and to download the plug-in, please visit *research.nokia.com/projects/activity_monitor*.

In order to appreciate how these sensors may be utilized for gameplay let us consider the Nokia 5500's 6g accelerometer (i.e., it can detect acceleration forces with a magnitude of up to six times that of earth's gravity). This accelerometer outputs three 12-bit, signed data values at a frequency of around 37 Hz.[11] These outputs correspond to the three phone axes (x, y, z) shown in Figure 6.9.

In terms of illustrating the opportunities for creating novel interaction methods using phones with inbuilt accelerometers, it is worth considering the following three Figures (6.10, 6.12, 6.13) which show the recorded output from the sensors on the Nokia 5500 in three different scenarios. Note that output has been expressed as acceleration forces in g for clarity.

Figure 6.10 shows the three accelerometer outputs when the phone was static on a desk (representing a horizontal plane). It can be seen that

---

[11] Vajk T., Bamford W., Coulton P., and Edwards R., *Using a Mobile Phone as a Wii-like Controller*, The Third International Conference on Games Research and Development 2007 (CyberGames 2007), Manchester, UK, 10–11 September 2007.

Figure 6.9   Nokia 5500 3D motion axes

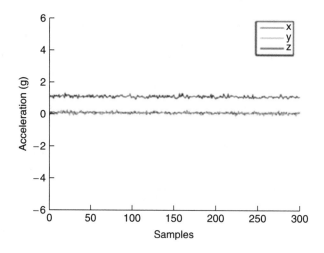

Figure 6.10   Nokia 5500 accelerometer data (phone at rest on a table)

outputs x and y are approximately zero (although some sensor noise is evident), and output z is showing a positive 1*g* force, which is the effect of gravity on the device. Note that this noise could be reduced by the application of a digital filter on the output values, but, in this case, we are showing the raw data. Overall, this figure illustrates that gravity provides a means of deducing the orientation of the phone which can then be utilized to provide a 'tilt' based controller.

In terms of games implemented using tilt on mobile phones, there have been numerous implementations of the old marble rolling games using

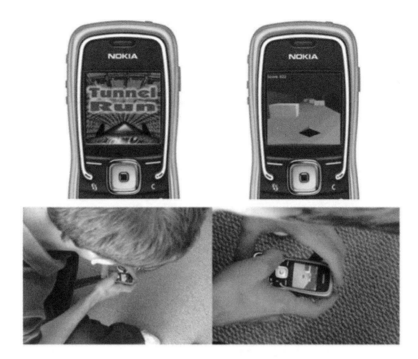

Figure 6.11 *Tunnel Run* tilt controlled driving game

the camera and one for the Nokia 5500 using the inbuilt accelerometers. Whilst these games are no doubt fun, they are replicating the existing game input mechanism of previous games, and one game, thus far, has explored tilt in relation to other game genres. This is the 3D graphics first-person driving game called *Tunnel Run* shown in Figure 6.11.[12]

In Figure 6.12, we see the accelerometer outputs resulting from several rapid lateral movements of the phone on a desk. At the start of the graph, we see outputs in the same state as in Figure 6.10, and then very large acceleration forces on the y axis produced by the movement. Please note, the output moves quickly between positive, then negative data values, indicating the transition between acceleration and deceleration and vice versa.

There are some smaller residual artifacts from this movement, seen in both x and z, which results from testing by hand, rather than using a fixed jig (as it is difficult to isolate a movement completely under such conditions). Figure 6.12 depicts the potential for using 'hand gestures' for the control of events within games and without the problems associated with using the camera to obtain the movement.[10] However, gesture recognition requires careful study as the variation in how the user holds

---

[12] Gilbertson P., Coulton P., and Vajk T., *Using Tilt as the Input for 3D Mobile Games*, The Third International Conference on Games Research and Development 2007 (CyberGames 2007), Manchester, UK, 10–11 September 2007.

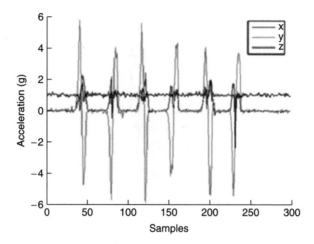

Figure 6.12 Nokia 5500 accelerometer data (lateral movement of phone across the table)

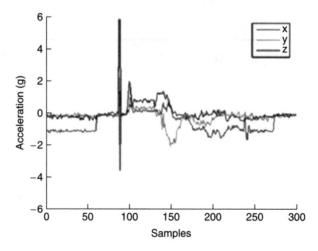

Figure 6.13 Nokia 5500 accelerometer data (phone dropped)

the phone could produce anomalous outputs and there is no method of obtaining the phone's physical position within actual space as provided by the sensor bar for the Wii. This is because both rotation and translational acceleration affect the accelerometer's output (essentially they can produce the same internal forces on the accelerometer).

Figure 6.13 represents the output after the phone has been dropped. The trace shows that initially the phone is held upright with the screen held at the top and gravity is acting on x. When the phone is dropped, the effect of gravity is overcome, and all three accelerometers output approximately zero before the phone hits the floor with a jolt. The aim of this test was to show that the phone could be used to measure activity of

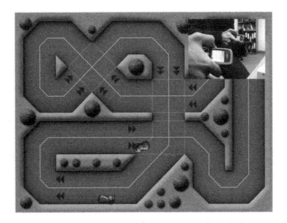

Figure 6.14 A screenshot from *Tilt Racer* and an example of typical play (inset)

a player within a game, such as jumping or running, whereby the phone would simply be worn rather than held and directly controlled.

Whilst there are undoubtedly many applications of 3D motion sensors that can be created using this novel phone interface, they could also be used to allow the phones to act as controllers for games running on large public displays as shown by *Tilt Racer* in Figure 6.14. *Tilt Racer* is a novel multiplayer game developed using the Microsoft XNA framework running on PC and displayed on a large public display.[10] The players' cars are controlled using Bluetooth technology.

6.4.1   Using Sensors on Symbian OS

Accessing the 3D motion sensors requires the use of the Symbian OS Sensor API which provides access to a wide range of sensors. These sensors can include accelerometers, thermometers, barometers, and humidity monitors. In fact, it can be any type of sensor designed to be incorporated into a mobile phone or be accessible to the phone via Bluetooth technology.[13] Sensors need only be supported by the API library to be usable. Currently, the Symbian OS Sensor API is available only from Nokia and requires the use of the S60 3rd Edition SDK, and, as such, this section is currently biased towards S60, although we would expect other handset manufacturers to follow suit in due course.

The API contains a number of key classes for retrieving sensor information and these provide the main methods for reading sensor data through callback functions. They are defined in `RRSensorApi.h`, and the application must link against `RRSensorApi.lib`.

---

[13] Coulton P., Bamford W., Chehimi F., Gilbertson P., and Rashid, O. *Using In-built RFID/NFC, Cameras, and 3D Accelerometers as Mobile Phone Sensors*, in *Mobile Phone Programming [0]and its Application to Wireless Networking* edited by Frank H. P. Fitzek and Frank Reichert, Springer 2007 (ISBN 1402059681), pp 381–396.

- `CRRSensorApi` is the sensor manager class that manipulates the available sensors on the phone.

- `MRRSensorDataListener` is the interface that must be implemented by the class to receive and interpret sensor data and it contains only one function `HandleDataEventL()`. This function takes two arguments: `TRRSensorInfo` and `TRRSensorEvent`.

- `TRRSensorInfo` is the class that holds the sensor information. It contains a human readable sensor name, a unique identifier for the sensor, and a category that identifies if it is internal to the device or external (over Bluetooth).

- `TRRSensorEvent` is the class that contains the data obtained from the sensor. The current implementation provides three `TInt` fields that contain the data. Each sensor will provide different data in these fields; some may provide absolute values, while others may provide values relative to sensor events.

Sensor API Overview	
Library to link against	`RRSensorApi.lib`
Header to include	`RRSensorApi.h`
Required platform security capabilities	`None`
Classes implemented	`MRRSensorDataListener`
Classes used	`CRRSensorApi`
	`TRRSensorInfo`
	`TRRSensorEvent`

Since the sensor APIs are manufacturer-dependent, as with vibration, and the Nokia 5500 is the only Symbian smartphone in the European market with a 3D sensor installed (at least at the time of writing), the code example we provide here for the test application, **TiltMe**, is therefore tested only on that phone.[14] The example shown in Figure 6.15 provides

Figure 6.15   Screenshots of **TiltMe**

---

[14] Note that testing this example can only be done on Nokia 5500 hardware, as there is currently no Windows emulator available for the sensors.

a simple text and graphical representation of the phone's tilting position in 3D space.

As with our previous examples the sensor setup is performed in the view class of the Symbian OS application framework. In the header file declaration of class `CTiltMeAppView`, we perform the following steps as shown in the subsequent code:

- include `RRSensorApi.h`

- inherit from the interface `MRRSensorDataListener`

- implement the interface's only pure virtual function `HandleData-EventL()`

- instantiate the sensor object `iAccSensor`, and some variables

- declare the start and stop functions of the sensor.

```
#include <RRSensorApi.h>

class CTiltMeAppView : public CCoeControl, MRRSensorDataListener
  {
public:
  // From MRRSensorDataListener
  void HandleDataEventL( TRRSensorInfo, TRRSensorEvent );
  // Starting the use of the sensor
  void StartSensor();
  // Stopping the use of the sensor
  void StopSensor();
private:
  // Accelerometer sensor
  CRRSensorApi* iAccSensor;
  // Time calculating variables
  TTime iFirst;
  TTime iSecond;
  // Sensor status
  TBool iSensorFound;
  // The X,Y,Z acceleration values
  TInt iXAccl;
  TInt iYAccl;
  TInt iZAccl;
  };
```

The two functions, `StartSensor()` and `StopSensor()` declared in `TiltMeView.h`, are used to control the 3D sensor. The user starts the sensor by simply clicking the 'Start' item on the 'Options' menu of the application. This will trigger a call to `StartSensor()` which constructs and registers `iAccSensor`. But, before we can use the sensor hardware, we must enumerate all the sensors on the device, using function `CRRSensorApi::FindSensorsL()`, and select the one needed, as there is a possibility of having more than one sensor installed. The resulting list of sensor(s) is stored on a `TRRSensorInfo` array as follows:

```
RArray<TRRSensorInfo> sensorList;
CRRSensorApi::FindSensorsL(sensorList);
```

Once this list is available, we need to iterate through it to find the sensor with the ID that matches the Nokia 5500 accelerometer sensor ID: 0x10273024. Note that this is a preconfigured implementation value set for the sensor hardware, and it is expected to be the same in future phones, to allow back/forward-compatibility. When a match is successful, `iAcc-Sensor` is constructed using the static factory function `CRRSensor-Api::NewL(TRRSensorInfo aSensor)`, with `aSensor` being the matching sensor found. The sensor is then registered by making a call to `CRRSensorApi::AddDataListener(MRRSensorDataListener* aObserver)`, where `aObserver` is a pointer to the class implementing the `MRRSensorDataListener`.

The following code snippet demonstrates the sequence described above:

```
void CTiltMeAppView::StartSensor()
  {
  // If the sensor is initially not found
  if( !iSensorFound )
  {// Get list of available sensors (if any)
  RArray<TRRSensorInfo> sensorsList;
  CRRSensorApi::FindSensorsL(sensorsList);

  // Number of sensors available
  TInt count = sensorsList.Count();

  for( TInt i = 0 ; i != count ; ++i )
    {
    // If the ID matches Nokia 5500
    if( sensorsList[i].iSensorId == 0x10273024 )
      {
      // Use the sensor found
      iAccSensor = CRRSensorApi::NewL( sensorsList[i] );
      // Register the sensor for our app
      iAccSensor->AddDataListener( this );
      iSensorFound = ETrue;
      return;
      }
    }
  // Omitted for clarity
  }
```

A call to function `CRRSensorApi::RemoveDataListener()` stops the sensor object from obtaining acceleration information from the sensor server. This is implemented in function `StopSensor()` as follows:

```
void CTiltMeAppView::StopSensor()
  {// If the sensor has been registered
  if( iSensorFound )
```

```
    {
    iAccSensor->RemoveDataListener();
    iSensorFound = EFalse;
    // Just to clean up behind us
    delete iAccSensor;
    iAccSensor = NULL;
    }
  }
```

Function `HandleDataEventL()` is the only pure virtual function to implement for the mixin `MRRSensorDataListener`. It receives callback notifications from the 3D sensor server and stores them in the acceleration values `iXAccl`, `iYAccl`, and `iZAccl` of our application, which are then printed on screen. In our example, these two operations are performed 10 times a second, and the timing is controlled by the two `TTime` member variables, `iFirst` and `iSecond`, whose microsecond interval is checked every time new acceleration data from the server is available:

```
void CTiltMeAppView::HandleDataEventL(TRRSensorInfo aSensor,
                                      TRRSensorEvent aEvent)
  {
  iSecond.HomeTime();
  // Find the interval from the second time member var to the first.
  TTimeIntervalMicroSeconds interv = iSecond.MicroSecondsFrom(iFirst);
  TInt64 i = interv.Int64();
  if( i >= 100000 ) // 1/10th of a second
    {
    iXAccl = aEvent.iSensorData1;
    iYAccl = aEvent.iSensorData2;
    iZAccl = aEvent.iSensorData3;
    // Reset the first time to the current one.
    iFirst.HomeTime();
    // Update the screen
    DrawNow();
    }
  }
```

Updating the screen is performed by a call to `DrawNow()`, which calls `Draw()` to display a compass-like needle arrow from the values of `iXAccl`, `iYAccl`, and `iZAccl`. The needle points at the direction of the tilting of the phone with its length indicating the tilting magnitude.

6.5 Vibration

Mobile phones have provided vibration since the very early models, as a means of increasing the likelihood of the user observing they are receiving a call or message, or for use in situations where sound is inappropriate. At first glance, the use of vibration may not immediately seem significant to developers, although one could imagine it being used for alerting users in games not requiring constant interaction. In fact, in the console and PC game industry, there has been a growing trend towards the use of haptic interface technologies (tactile feedback such as vibration), and such interactions are equally relevant for mobile games.

Haptic technologies interface to the player through their sense of touch by applying vibration and/or motion through the game controller. In most cases, this interface creates the illusion of forces acting on the player, as in reality, true force feedback would involve the feedback of an actual resisting force to the player. A common example of this type of haptic feedback is in driving games, where, by the simulated revolution of the engine, the imaginary bumps on the road are transferred to the user through vibrations of the controller, to enable greater immersion within the game.

Thus far, the use of vibration in mobile games has been limited, but there are products appearing that may expand its use and make it easier for developers to create more sophisticated uses for it. Most notable is VibeTonz from Immersion, which offers a combined software and hardware solution in the form of a dedicated player and amplifier design, which has been licensed by Samsung and Nokia with commercial games already appearing.

6.5.1   Using Vibration on Symbian OS

Controlling the vibration motor in phones via Symbian OS was not a supported feature for third-party developers in the early releases of the different Symbian OS-based UI platforms. The APIs were private components, meaning that they were restricted to the licensees of Symbian OS and their partner companies. The reason for this was, most likely, that vibration consumes a lot of battery power, and manufacturers wanted to limit excessive demands on this constrained resource.

Advances in hardware and power supply technologies meant that, from Symbian OS v7.0, these APIs became public. However, they were not always consistent and they were sometimes missing from different firmware builds of the same phone model. In addition, there were backward-compatibility issues, even between different platforms from the same manufacturer, which caused portability problems for developers. For example, applications developed for S60 2nd Edition Feature Pack 2 (FP2) devices (such as the Nokia 6630) that implemented the vibration feature used to crash at run time on S60 2nd Edition FP1 phones (such as the Nokia 3230). Consequently, the use of vibration was not recommended when developing applications that were targeted to run on handsets spanning several platform versions. This limitation has since been resolved in recent releases of the S60 3rd Edition SDK, because Nokia has deprecated the old vibration APIs CVibraControl (the interface to control the vibration motor) and `VibraFactory` (which created instances of `CVibraControl`). Nokia is in the process of phasing those classes out entirely and replacing them with a new API `CHWRMVibra`, which is what you should now use for S60 3rd Edition devices.

The following example shows the usage of `CHWRMVibra` for the Nokia S60 3rd Edition platform, as it is the one to be supported in future phones by Nokia, and it does not introduce any compatibility breaks.

Vibration API Overview	
Library to link against	`HWRMVibraClient.lib`
Header to include	`hwrmvibra.h`
Required platform security capabilities	`None`
Classes to implement	`MHWRMVibraObserver` (dependent)
Classes to use	`CHWRMVibra`

As a simple example, we have created the **_VibraPool_** game shown in Figure 6.16. The application demonstrates how vibration could be used as part of a pool game where collisions of the ball on the sides of the pool table are indicated by short vibrations, while longer ones indicate the ball has gone into a pocket.

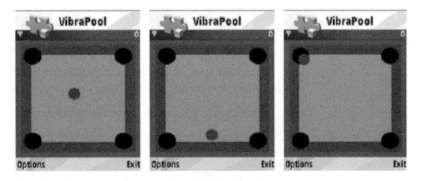

Figure 6.16 Screenshots of **_VibraPool_**

Using the vibration feature in S60 is a fairly straightforward task and simply requires the instantiation of an object of class `CHWRMVibra` using the static factory function `CHWRMVibra::NewL()`. It is important to note that the `NewL()` function may leave with `KErrNotSupported` if the phone does not support the vibration feature, or `KErrNoMemory` if insufficient memory is available to instantiate the `CHWRMVibra` object.

The vibration is started by calling the function `CHWRMVibra::StartVibraL()`, passing in a specific duration measured in milliseconds. A value of 0 specifies that the vibration should continue indefinitely until it is stopped by a call to `CHWRMVibra::StopVibraL()`. Note that the duration can alternatively have a maximum value of `KHWRMVibraMaxDuration` (=2147482).

`StartVibraL()` has another overload that takes intensity as a second parameter (a value between 100 and −100) to represent the percentage of the rotation speed of the vibration motor. Note that the device might have hardware imposed limits on the supported vibration intensity values, so actual effects might vary between different hardware. A negative value for intensity rotates the vibration motor in one direction, while a positive value rotates it in the opposite direction – each feels different to the touch. Value 0 stops the vibration altogether.

Either of the `StartVibraL()` overloads may leave with one of the following error codes as documented in the S60 SDK:

- `KErrNotSupported` if the device doesn't support user-defined vibration intensity (specific only to the second overload of `StartVibraL()`

- `KErrArgument` if the duration and/or intensity are out of range

- `KErrAccessDenied` if the vibration setting in the user profile is not enabled

- `KErrBadHandle` if the vibration session has been invalidated (see SDK documentation)

- `KErrLocked` if vibration is prevented by a lock down because of too much continuous use, or if it is explicity blocked by, for example, some vibration sensitive accessory

- `KErrTimedOut` if a timeout occurred in controlling the vibration

- `KErrInUse` if using vibration is not reserved to this client but it is reserved to some other client

- `KErrNoMemory` if there is a memory allocation failure

- `KErrGeneral` if there is a hardware error.

As the vibration feature on the phone can also be adjusted by the user, the different mode settings and status values can be queried by calling `CHWRMVibra::VibraSettings()` and `CHWRMVibra::VibraStatus()` respectively. The modes are specific to the vibration settings in the user profile and they are enumerated in the class `TVibraModeState` with the following possible values:

- `EVibraModeON` if the vibration setting in the user profile is on

- `EVibraModeOFF` if the vibration setting in the user profile is off

- `EVibraModeUnknown` if the vibration is not initialized yet or there is an error condition.

The vibration status is responsible for querying the current status set for the vibration by this application or another client. It is defined by the `TVibraStatus` class and can be:

- `EVibraStatusUnknown` if the vibration is not initialized or because of an uncertain error condition

- `EVibraStatusNotAllowed` if the vibration is not enabled in the user profile, or some application is specifically blocking it

- `EVibraStatusStopped` if the vibration is stopped

- `EVibraStatusOn` if the vibration is turned on.

Changes to the phone's vibration mode or status generate notifications to implementers of the mixin class `MHWRMVibraObserver`. It provides two pure virtual functions, `VibraModeChanged()` and `VibraStatusChanged()`, that must be implemented by any class wishing to receive change notifications. The programmer should implement this observer class if handling the vibration state events matters for the application (but it is not mandatory to do so). To receive change notifications, use the second overload of `NewL()`, which takes a reference to `MHWRMVibraObserver` as a parameter.

There are additional features available, relating to priorities and suspensions, although these are generally only required by very sophisticated applications and are unlikely to be required for games. The mechanism guarantees a certain client to have a higher priority to use the vibration control over other applications by reserving it using the `CHWRMVibra::ReserveVibraL()` function. As a result, other clients with lower priorities are forced to suspend their use of the resource temporarily until it is released by `CHWRMVibra::ReleaseVibra()`.

Having provided an overview of the API we can apply it to our example starting from our `CVibraPoolView`. We start with `VibraPoolView.h` and perform the following actions in the subsequent code:

- include `hwrmvibra.h`. It requires linking against `HWRMVibraClient.lib` in the MMP file

- declare the vibration instance `iVibra`
- declare `iMove` to animate the ball
- declare the controlling functions of the game.

```
#include <hwrmvibra.h>  // Vibration API

class CVibraPoolView : public CCoeControl
    {
public:
  // Construction, destruction omitted for clarity
public:
  void ReStartGame();// Restarts ball animation
  void StopGame();// Stops ball animation
private:
  // Game action functions
  void WallsDetection();
  TBool HolesDetection();
  void UpdateScreen();// Updates new ball
  // position on screen
private:
  CHWRMVibra* iVibra;
  CMoveBall* iMove;
  // Ball positioning variables omitted for clarity
    };
```

The construction of `iVibra` and `iMove` is performed in `CVibraPoolView::ConstructL()`, which also initializes the values for the ball position and movement on the table. The logic behind moving the ball and detecting walls and holes is not the main aim of the example, and, therefore, we will cover only the vibration implementation details; please download the full sample code for the application from the Symbian Developer Network (***developer.symbian.com/gamesbook***) for more details.

```
void CVibraPoolAppView::ConstructL( const TRect& aRect )
    {
  // Constructing the vibration control
  iVibra = CHWRMVibra::NewL(); // No callbacks
  // Initial ball location on screen (set randomly)
  iBallLoc.iX = Math::Random()%
                ( aRect.Width()-(TABLE_BORDER*2)-
                  BALL_DIAMETER-(THRESHOLD*2) )+
                ( THRESHOLD+TABLE_BORDER );
  iBallLoc.iY = Math::Random()%
                ( aRect.Height()- (TABLE_BORDER*2)-
                  BALL_DIAMETER-(THRESHOLD*2) )+
                ( THRESHOLD+TABLE_BORDER );

  // Number of pixels to step each time the ball moves.
  iMoveX = iMoveY = STEP;
  // Ball not in a hole so it is drawn
  iDrawBall = ETrue;
  // Construct and start our asynchronous timer to move the ball.
```

```
iMove = CMoveBall::NewL(*this);
iMove->StartMovingBall();
}
```

Up to this point, we have created the framework such that the vibration control is ready to use and now we just need to start or stop it as required. Function `WallsDetection()` uses vibration in two different scenarios. It first checks if the ball has hit the edge of the pool table, which it does by calculating the distance between the upper-left corner of the rectangle surrounding the ball, and any of the table edges. If it is less than the diameter of the ball (25) plus the width of the table edge (30) then it is a hit and `WallsDetection()` will generate a short vibration effect. The effect is generated by the function, calling `CHWRMVibra::StartVibraL()` to run the vibration motor for an interval of 100 milliseconds and an intensity of 20 %. The second scenario is applied when the ball falls into any of the holes. The effect for this scenario is longer than the border collision; it has two interruption intervals that switch the direction of rotation, and two intensities to form a rhythm-like vibration effect.

The following code illustrates `WallsDetection()` implementation for these vibration effects. Note that the function keeps checking for the mode of the vibration setting in the user profile. If it is on (`EVibraModeON`), the vibration will be initiated, otherwise only the ball movement will be in effect.

```
void CVibraPoolAppView::WallsDetection()
  {
  if(iBallLoc.iX+iMoveX >=
    (Rect().Width()-(TABLE_BORDER+BALL_DIAMETER-STEP)))
    {
    // Change the direction of the movement
    iMoveX = -STEP;
    // If vibration is on in the profile settings
    if(iVibra->VibraSettings() == CHWRMVibra::EVibraModeON)
      {
      iVibra->StartVibraL(100, 20);
      }
    }

  // After ball location is updated check if the ball
  // is to fall in a hole.
  if( HolesDetection() )
    {
    if(iVibra->VibraSettings() == CHWRMVibra::EVibraModeON)
      {
      // Vibrate in negative direction
      iVibra->StartVibraL(300, -100);
      // Update the screen now
      DrawNow();
      // First interval of half a second
      User::After(500000);
      // Vibrate in positive direction
      iVibra->StartVibraL(300, 100);
```

```
// Second interval of another half second
User::After(500000);
iVibra->StartVibraL(1000, -100);}
}
}
```

Overall, we would encourage developers to think of using vibration both in terms of its more obvious use (to enhance driving, boxing, or fighting games), and also less obvious implementations, such as alerting users to the presence of objects or other players in location games. Finally, it is important to consider that, although developers often consider sounds as a primary source of feedback within a game, mobile games are often played in public places, where the use of sounds may be inappropriate. Vibrations can be an important, and useful, alternative and should therefore be given due consideration.

6.5.2   Proximity/Presence

As we discussed previously in section 6.3, games may use proximity, in the sense that the user's location is close to either another player or a real or virtual artifact within the game. However, none rely solely upon the proximity between either a player and another player, or the player and a game artifact, irrespective of the physical location. The most notable example of this type of game playing, which is not mobile phone based, is **Pirates**,[15] which was an adventure game using personal digital assistants (PDAs) and RF proximity detection. Players completed piratical missions by interacting with other players and game artifacts using RF proximity detection. One of the interesting aspects that emerged was the stimulated and spontaneous social interaction between the players. An interesting prospect for the mobile game industry is to produce games that provide opportunities to utilize interaction of a more serendipitous nature for their players. We believe this serendipity can be best achieved by removing the requirement for a central game server, as utilized in **Pirates**, and using a proximity detection scheme that initiates a dynamic peer-to-peer connection between the mobile users when they enter into the close vicinity of each other, requiring no pre-arrangement for the meeting.

In terms of proximity detection, the obvious choice is Bluetooth-which, despite previous predictions of its demise, is in fact increasing its growth. In fact, there are already a small number of mobile Bluetooth proximity applications which are often described as mobile social software (MoSoSo) and can be viewed as evolutions of Bluejacking. This is a phenomenon where people exploit the contacts feature on their mobile

---

[15] Björk S., Falk J., Hansson R., and Ljungstrand P., *Pirates! Using the physical world as a game board*, Proceedings of Interact 2001, IFIP TC.13 Conference on Human-Computer Interaction, Tokyo, 2000.

phone to send messages to other Bluetooth-enabled devices in their proximity. Bluejacking evolved into dedicated software applications such as **Mobiluck** and **Nokia Sensor** which provided a simpler interface, and in the case of **Nokia Sensor**, individual profiles that could be used to initiate a social introduction. More complex systems such as **Serendipity**[16] have tried to introduce a greater degree of selection into the process in that a server tries to match users before making the introduction. There are relatively few games that have exploited Bluetooth in this way, but one such is **You Who** from Age0+ that provides a simple game premise to help initiate a meeting. After scanning for other users running the application and 'inviting' a person to play the game, the first player acts as a 'mystery person,' who then provides clues about their appearance to the second player, who builds up a picture on their mobile phone screen. After a set number of clues have been given, the players' phones alert, revealing both players' locations and identities. Obviously, the gameplay is quite limited and effectively non-competitive, which is unlikely to result in repeated game playing, therefore, it is closer to the Nokia Sensor than a game.

The game described above provides significant opportunities for serendipitous social interactions, addictive, and competitive game play. The game draws from the familiar, which is always a good way of gaining acceptance for a new game, using the concept of a wild west quick-draw gunfighter, which we have called **Mobslinger**,[17] as shown in Figure 6.17.

Figure 6.17   Screenshots of **Mobslinger**

[16] Eagle N., and Pentland A., *Social Serendipity: Mobilizing Social Software*, IEEE Pervasive Computing, Special Issue: The Smart Phone. April–June 2005, pp 28–34.

[17] Clemson H., Coulton P., and Edwards R., *A Serendipitous Mobile Game*, Proceedings of The Fourth Annual International Conference in Computer Game Design and Technology, Liverpool, 15–16 November 2006, pp 130–134.

Mobslinger runs as a background application on S60 smartphones and periodically scans for other users in the vicinity who are also running the *Mobslinger* application. Once detected, a countdown timer is initiated on both phones, which alerts the users by sounding an alarm and vibrating the phone. The user then has to 'draw' their phone and enter the randomly generated number which has appeared on the screen as quickly as possible. The person with the fastest time is the winner and the loser is 'killed,' which means their application is locked out from game playing for a set period of time.

6.6   Summary

In the past, mobile games have suffered from a reputation of poor functionality and usability, and have been criticized by some game developers, who have been unduly harsh about an industry which only really started in 1997 with the *Snake* game that was embedded in Nokia phones. However, this criticism is diminishing, and, as the mobile games industry comes of age, it is creating a vibrant market in its own right, built around the inherent strengths of mobile communications technology. In this chapter, we have highlighted just some of these strengths that give mobile game developers opportunities to create new game experiences that are uniquely mobile. There are others worthy of consideration, for instance, on a device with such powerful audio functionality, we haven't yet seen any audio driven games or interfaces.

While there are no doubt challenges in the commercial realization of some of these new mobile game experiences, we believe that, if embraced, mobile game developers can help establish the very innovative industry from under the shadow of the PC and console game market and play a significant role in the world of game development.

6.7   Acknowledgements

We would like to express our sincere thanks to Nokia and Symbian for support given to the Mobile Radicals research group within Infolab21 at Lancaster University UK, where many of the innovative games featured in this chapter were developed.

Part Three

Porting Games to Symbian OS

7

C/C++ Standards Support for Games Developers on Symbian OS

Sam Mason
(Mobile Intelligence)

7.1 P.I.P.S. Is POSIX on Symbian OS

P.I.P.S. is the result of a massive effort to make Symbian OS POSIX compliant (adhering to the ISO/IEC 9945 specification) and to open the doors to C programmers worldwide. It may sound like a cocktail, but P.I.P.S. is actually a recursive acronym: P.I.P.S. Is POSIX on Symbian OS.

Symbian OS was never originally designed to be POSIX compliant, but increases in hardware performance and advances in the operating system over the last few years (specifically the introduction of EKA2) means that it is now feasible to provide a platform for porting existing C applications to run on Symbian OS, and encourage the re-use of open source code originally developed on other operating systems. With the first release in January 2007, the P.I.P.S. initiative allows developers with desktop or server skills from Linux, Unix and Windows backgrounds, who program using the C language, to port and mobilize their games with relative ease. Although cited by Symbian as enabling the porting of popular desktop middleware and applications such as web servers and file sharing software, P.I.P.S. is also good news for games developers who use C.

That all reads well and makes a nice press release, but it doesn't mean a whole lot if you don't know what POSIX is in the first place, so let's cover that briefly first. Back in the dark ages, around about 1988, it was clear that there was a growing need to define a set of standard interfaces which would allow software to run unchanged on any system that supported the standard, and so the Portable Operating System Interface for Unix (POSIX) was born. Among other things, it defines requirements for threads, file and network I/O, security and IPC, as well as conformance testing.

Most Unix-like systems are basically either fully compliant (for example, Mac OS X, Solaris, AIX) or 'mostly' compliant (like FreeBSD and Linux) – which means that while not actually certified, they pretty much cover it all. Others, like P.I.P.S., are 'mostly' compliant by virtue of updates or add-ons to the operating system. P.I.P.S. is not a Unix virtual machine and you can't expect to just run all of your games all of a sudden – it's simply an extension to the standard operating system that supports ANSI C programming on Symbian OS.

In practical terms, what this means to the development community is a set of headers, libraries and a tool update in your SDK. And you can't use the P.I.P.S. libraries on the phones if they aren't there, so there are also signed SIS files to upgrade S60 and UIQ phones. On S60 3rd Edition, you can embed the P.I.P.S. runtime libraries into your own SIS file, by editing your package definition file as shown below:

```
@"pips_s60_1_1.sis", (0x20009A80)
```

And on UIQ 3:

```
@"pips_uiq_1_1.sis", (0x20009A81)
```

P.I.P.S is only available in Symbian OS v9.1 and later. From Symbian OS v9.3 onwards, the libraries will be included in the ROM, so built directly into the phone handset, with no need for separate installation using a SIS file.

At time of writing, the last P.I.P.S. release (version 1.1) was in July 2007. This provided P.I.P.S. for Symbian OS v9.1 but, as yet, there is nothing for v9.2 (due to a difference in the low-level drivers), but the next P.I.P.S. release will work for Symbian OS v9.2. P.I.P.S. version 1.1 will thus only work with the S60 MR 3rd Edition SDK and the UIQ 3.0 SDK. It also breaks binary compatibility with the previous version of P.I.P.S., which, while a pain, was necessary, because in the earlier version Nokia and Symbian were using different UID ranges for the libraries (see section 7.2 on Open C).

There are four main P.I.P.S. libraries:

- libc – standard C and POSIX APIs (see below)

- libm – mathematical APIs

- libpthread – thread creation and synchronization mechanisms

- libdl – standard C dynamic loading and symbol lookup APIs.

By including the libc library, P.I.P.S. provides Symbian OS with standard libraries including stdio, fileio and stdlib, as well as string manipulation, searching, sorting and pattern matching functionality, locale and systems services, networking and socket APIs, and IPC mechanisms (pipe, FIFO, message queues and shared memory). It also includes process creation methods, such as `popen`, `posix_spawn` and `system`.

At this point, you might be thinking of the standard library (STDLIB) that Symbian OS has included for ages (used originally to allow a JVM to run in Psion days). Well that's now scheduled to be deprecated, so don't use it any more – it doesn't support threading and it isn't compatible with P.I.P.S. anyway, as it was designed in a different time with a completely different set of aims in mind.

In terms of porting games across to Symbian OS using P.I.P.S., there are a number of issues that need to be considered. To start with, notice that there are no graphics libraries yet, so any game with a widget-based GUI will need to have its interface completely re-written to use S60 or UIQ controls. But as there's no doubt a clean separation between UI and engine code in the existing C system, that won't be a problem, right?

Further, Symbian OS doesn't support symbolic lookup in dynamic libraries in any version prior to v9.3. Two new target types were created in Symbian OS v9.3 to flag this standard usage (`STDDLL` and `STDEXE`) and set some MMP defaults (like enabling writable static data), but, more importantly, the loader was changed to support both ordinal and symbolic lookup in DLLs. This allows symbolic linking to be supported from Symbian OS v9.3 in P.I.P.S. too. The workaround for versions earlier than Symbian OS v9.3 is simply to find out the ordinal number for the function from the DEF file and pass it into the `dlsym` method from the `libdl` library:

```
// Instead of
dlsym(&handle, "CreateWanderingMonster");

// If CreateWanderingMonster() has ordinal 3, use
dlsym(&handle, "3");
```

There are some additional considerations that arise because P.I.P.S. must obey the same rules of Symbian OS platform security as other native applications. For example, platform security prevents applications from reading or writing to each other's data caged areas without the `AllFiles` capability (which it's unlikely a game should need or get). So a game that writes data to local temporary files needs to be aware that its default will be its data caged area. Likewise, sharing a file descriptor between processes (used in POSIX for IPC) will fail without the correct capabilities for the location of the file in question. Another example is the use of sockets, which requires the `NetworkServices` capability. For more information about platform security, you should consult the Symbian OS library documentation inside the SDK you've downloaded for the platform you're working on (UIQ or S60). There is also a comprehensive treatment in the Symbian Press book called *Symbian OS Platform Security*, by Craig Heath, published in 2006.

On POSIX-compliant systems, a process can use `fork` or `exec` to create, manage and communicate with child processes. However there is no support for this on Symbian OS even with P.I.P.S., although there are a number of situation-specific workarounds that can be used where necessary. Depending on the game, this may or may not be a major problem.

At the time of writing, P.I.P.S. doesn't have any support for signals and, as a consequence, game code that is designed to operate with either signal or exception handlers should be removed. The main reason signals are not supported at present is that P.I.P.S. implements many POSIX system calls using standard IPC mechanisms on Symbian OS. These methods aren't re-entrant, as they were never designed to be interrupted by signals in the first place.

A game port will also need to address any mechanisms that use signals for IPC by changing these to use pipes or FIFOs and would also need to move all asynchronous I/O (which also isn't supported by P.I.P.S. because, in the absence of signals, all I/O must be synchronous) into separate threads in the same process. Note, however, that this is within a P.I.P.S. context only – you can obviously still perform asynchronous I/O using native Symbian OS C++ calls, as and when required.

With each subsequent release of P.I.P.S., Symbian OS is becoming closer to being a fully POSIX compliant operating system as per the IEEE standard. And hopefully as P.I.P.S. matures, we'll see a wide range of exciting game titles ported in the future.

7.2 Open C

This is Nokia's answer to the same problem addressed by P.I.P.S. – Nokia has taken it further by adding another five standard C libraries on top of the four provided by P.I.P.S. Open C is only available for S60 3rd Edition devices. On UIQ 3, only the four P.I.P.S. libraries (effectively a subset of Open C) are available.

Symbian and Nokia have been working together very closely over the last 18 months to keep both Open C and P.I.P.S. in line with each other. Open C is available as a plug-in to the S60 3rd Edition SDK (Maintenance Release) from Forum Nokia (although you can also use S60 3rd Edition FP1, but you can't target any handsets if you do). Installing the plug-in will update your tool chain and add the necessary libraries and headers to your file system. You can use Carbide.c++ 1.1/1.2, CodeWarrior 3.1 or Microsoft Visual Studio (with the Carbide.vs plug-in) for Open C development, although you will need Carbide v1.2 if you wish to take advantage of the new target types defined for Symbian OS v9.3 (`STDDLL` and `STDEXE`).

Table 7.1 Open C and P.I.P.S. libraries and their coverage

Standard Library	Functionality provided by	% Standard functionality implemented
libc	P.I.P.S. (UIQ and S60)	47
libm	P.I.P.S. (UIQ and S60)	42
libpthread	P.I.P.S. (UIQ and S60)	60
libdl	P.I.P.S. (UIQ and S60)	100
libz	Open C (S60 only)	100
libssl	Open C (S60 only)	86
libcrypt	Open C (S60 only)	100
libcrypto	Open C (S60 only)	77
libglib	Open C (S60 only)	77

Since many functions in the standard C libraries are rarely used in practice, not all of them have been included in the Open C versions in the interests of keeping memory footprints as small as possible. Table 7.1 shows the relative coverage as well as showing the five extra libraries that differentiate Open C from P.I.P.S. These include libraries for compression (libz), SSL and TLS (libssl), cryptography (libcrypt and libcrypto), and the general purpose utility library libglib, which provides string and file utilities, data types, and macros.

Restrictions on intellectual property rights prevented the implementation of some cryptographic algorithms (such as Blowfish, MD4 etc.), and, since there is no parent-child relationship in process spawning, some of the APIs in libglib are not supported.

Since Symbian smartphones to date haven't had floating point processors, floating point operations have been emulated in software and consequently floating point exceptions are not supported in Open C (although now there are several Nokia devices – including the Nokia N82, N93 and N95 – that do have floating point co-processors).

As with P.I.P.S., library symbolic lookup is not supported because under Symbian OS, DLL entry points are exported by ordinal only. We've already seen how to deal with this when using the dlsym function from libdl, and the analogous case holds when you use g_module_symbol from libglib:

```
// instead of
ret = g_module_symbol(module, "LaunchDrone",&ptr);

// where LaunchDrone() has ordinal 22, use
ret = g_module_symbol(module, "22",&ptr);
```

This is only necessary with the Open C SDK plug-in since Symbian OS v9.3 has added new target types (STDDLL and STDEXE), which introduce symbolic lookup support. This will be included in the release of the S60 3rd Edition SDK that is based on Symbian OS v9.3.

It will also add other features such as automatically including the Open C libraries during compilation, linking in euser.lib and defaulting the contents of the MMP file so that you don't have to specify the start up library. Note that native Symbian OS C++ DLLs and EXEs will not support lookup by name – this is specific to the new target types.

Again, you'll need to add runtime libraries on actual phones; to this end, the Open C plug-in comes with four SIS packages for use on hardware as described in Table 7.2.

Table 7.2 Open C SIS packages

Package	UID	Inclusion	Notes
pips_s60_wp.sis	0x20009A80	Mandatory	Contains libc, libm, libdl and libpthread
openc_ssl.sis	0x10281F35	Optional	Contains libcrypt, libcrypto, libssl and libz
openc_glib.sis	0x10281F2C	Optional	Contains libglib
stdioserver.sis	0x20009AA2	Optional	Supports a console to be used only for testing

When your game is built and ready to be deployed onto hardware, you embed the relevant SIS files into your own package, as shown below, and away you go.

Notice that, while the SIS file is named differently to that supplied with P.I.P.S. (pips_s60_1_1.sis), the UID remains the same.

```
@"pips_s60_wp.sis", (0x20009A80)
```

Open C has already met with some success, even though it was only released in January 2007. Oracle recently ported Berkley DB across to S60 in less than 10 person-days, requiring changes to only about 1000 lines of code,[1] and a quick browse through the discussion groups at Forum Nokia shows that C programmers are already trying out a wide variety of porting projects across many domains. This isn't really surprising when you consider that, for the first time, C programmers can now have a valuable part to play in the development of mobile applications and games – particularly considering the huge volume of OpenGL ES libraries written in C out there, which can now run almost unchanged under S60!

So the skills of all these game developers working on C can be focused – they can write or port AI libraries, graphics libraries, or even servers for local clients, all very easily using Open C. These are simply linked against and used just like any other library. The advantages to game ports here should be obvious as this mixture allows each set of experts to do what they do best. However, a developer will still need to have at least a nodding acquaintance with core Symbian OS concepts such as descriptors, leaves and the cleanup stack, and the Symbian OS tool chain when working on the platform. This isn't a particularly easy learning curve; it's often by-passed as questions on the developer forums would seem to indicate.[2] More than one person has jumped straight in and lost days trying to port across software that uses sockets over WLAN or GPRS without assigning the `LocalServices` or `NetworkServices` capabilities in their MMP file. Data caging seems to be another thing causing issues, as developers attempt to access file system locations that are protected and require certain capabilities.

In addition, when it comes to porting games using Open C there are some things that need to be kept in mind:

- there are no UI libraries available to date

- Open C doesn't provide access to telephony, messaging, Bluetooth technology or location-based services, or other APIs specific to Symbian OS

- as already mentioned, there is no support for `signal`, `fork` or `exec`, which is likely to be the biggest hurdle

- Open C libraries are still subject to the platform security restrictions, so games will still need to be assigned the relevant capabilities where they are required

- if you mix C and C++ code, you need to be careful to specify C APIs as such (using `extern "C"`) or their names will get mangled by the C++ compiler

---

[1] More information can be found at **www.nokia.com/NOKIA_COM_1/Developers/ Success_Stories/Enablers_&_infrastructure/Dev_succ_Oracle.pdf**.

[2] A good place to ask questions about Open C is on the discussion board at **discussion.forum.nokia.com/forum**. The P.I.P.S. forum at **developer.symbian.com/forum** is another good place to find information.

- while you can mix C code and Symbian OS C++ code, it's a good idea to keep this separate as much as possible because C does not have any concept of leaves or the cleanup stack. Allocating memory using C functions and then calling functions that may leave is asking for a memory leak (although you can use `TCleanupItem` for this, you'd have to ask why you're coding in this manner in the first place)

- Open C is called that for a reason – it doesn't help porting over of C++ games or applications. Currently, there's no official STL implementation.

So the take home message here is that Open C allows C programmers to port applications and games onto S60, but you'll more than likely need expertise in both Symbian OS and the C programming language to effect a truly successful transition.

Open C demonstrates Nokia's continuing commitment to working with Symbian to keep the world's leading smartphone platform as the best of breed. By providing support for industry standards such as POSIX compliance, we should see a large influx of game titles over the next year or so as the potential of this technology is realized in the market.

7.3   OpenKODE

As Aleks explained in Chapter 4, Symbian and Nokia are both members of the Khronos Group (**www.khronos.org**) which is *"a member-funded industry consortium focused on the creation of open standard, royalty-free APIs to enable the authoring and accelerated playback of dynamic media on a wide variety of platforms and devices."*

Some of their more relevant standards for games include COL-LADA (an XML based standard for content authoring), glFX (a visual effects abstraction framework) and, of course, our current subject OpenKODE.

To quote directly from Khronos group's presentation material, *"OpenKODE is a set of native APIs [that aims] to provide source portability."*

This initiative is specifically aimed at resource-constrained devices such as mobile phones. The obvious driving force here is the proliferation of hardware platforms and operating systems in the mobile space, of which Symbian OS is just one.

Consumer demand and game development economics are pushing the consolidation of standards in many areas. Enter OpenKODE, which is actually comprised of a suite of other open standards managed by the Khronos Group along with definitions to tie it all together. It includes hardware acceleration requirements for the following four standard media-centric APIs: OpenGL ES (OpenGL for embedded systems), OpenVG (vector graphics), OpenMAX (streaming media portability) and OpenSL ES (embedded audio).

It's often referred to as being just 'like DirectX for mobile phones' – except for the obvious caveats that it's royalty-free, cross-platform, designed for handheld devices and is an open standard (so really it's probably more accurate to say that it's not at all like Microsoft DirectX, other than the fact that they're both media-related umbrella technologies).

The main point about OpenKODE is that it defines two key features as part of its standard. First, it defines a set of APIs that serve as core operating system abstractions, which is clearly necessary for cross-platform support. This includes the following:

- core OS libraries based on the POSIX standard as far as possible

- thread creation and synchronization

- an event system with key OS abstractions (window resize, pause, quit, etc.)

- mathematical functions – mainly standard C analogs

- utility functions – assertions, logging, string conversion, memory allocation, random number generators (RNG) and so on.

- C time function analogs and some more accurate OpenKODE-specific time keeping functions for better accuracy, as well as event raising timers

- a virtual file system based on C and POSIX file I/O

- networking based on BSD (POSIX) sockets

- cryptographic support

- user input abstractions for key strokes (similar to the game keys used in Java MIDP) and access to outputs in a standard manner

- window support for multiple non-full screen windows, resize, minimizing, maximizing and so on.

Secondly, OpenKODE defines in no uncertain terms exactly how a wide variety of media APIs can interact with each other in a meaningful way to allow the creation of media-rich content and advanced UIs for mobile phone applications and games. This is done using EGL 1.0 as a kind of communication API; it is actually an abstraction of the underlying operating system. EGL is described later in section 7.7.

At the time of writing, the current version of the OpenKODE standard is the 1.0 Provisional Release, which contains OpenGL ES 1.1 and OpenVG. The first half of 2008 should see version 1.1 of the specification, which will also include the OpenSL ES and OpenMAX standard APIs.

As game developers, you may be asking why you care, but the truth is that on small screen devices, hardware accelerated media can add up to dramatic power savings (of about an order of magnitude when compared to software emulation). This saving is a major advantage in terms of phone battery life alone. As well as that, your animations and graphics will be that much smoother, and the processor cycles that aren't being used for media calculations can then be put to better use (for more complex physics simulations and game AIs which will make your game more immersive).

The games industry is already supporting the OpenKODE standard with Softbank, a major wireless operator in Japan. The Khronos group announced in May 2007[3] that Softbank will be incorporating support for OpenKODE into the Portable Open Platform Initiative (POP-i) which will support major mobile operating systems including (but not limited to) Linux, Windows Mobile and of course Symbian OS. Softbank became a member of the Khronos Group in August 2007.

Even better from a game developer's point of view is that Ideaworks3D and Texas Instruments (TI) announced in March 2007 that TI's OMAP™ Platform[4] will be extended to support the OpenKODE 1.0 specification using the Airplay SDK.[5] And even better than that, it will scale for future standards such as OpenGL ES 2.0 (which will have support for shaders, and frankly that's just going to be berserk).

Leon Clarke from Ideaworks3D has this to say about the Airplay SDK and OpenKODE:

"The goal of the joint project with TI and Ideaworks3D is that developers can start developing OpenKODE games using the Airplay SDK on top of TI's development boards, meaning that developers can start exploiting the features of new versions of OMAP before they are integrated into real phones. Because Airplay provides binary compatibility across a wide range of platforms and makes fanatical attempts to behave in the same way on all platforms, when phones eventually arrive with the new OMAP platform, developers can expect their games to just run on the new device, even if for some strange reason, the handset is running Windows Mobile or BREW instead of Symbian OS. Airplay also supports ARM's RTSM real-time simulator, allowing the actual ARM binary to be debugged in a source-level debugger on your desktop without even needing a device. There's also a Windows-native emulator allowing you to use the full capabilities of Visual Studio 2005's debugger.

---

[3] For more details, see the press release on the Khronos website: ***www.khronos.org/news/ press/releases/softbank_mobile_adopts_openkode_to_provide_rich_media_acceleration*** or a related article at ***www.linuxdevices.com/news/NS5874940517.html***.

[4] For more information about the OMAP Platform, please see: ***focus.ti.com/general/ docs/wtbu/wtbugencontent.tsp?templateId=6123&navigationId=11988&contentId=4638*** or search for OMAP from the main TI website at ***www.ti.com***.

[5] More information can be found in the Ideaworks3D press release at ***www3. ideaworks3d.com/downloads/Ideaworks3D-TI_OpenKODE.pdf***.

The approach taken by Ideaworks3D to OpenKODE is slightly unusual, in that it is built on top of Airplay. Ideaworks3D was very heavily involved in the development of the OpenKODE standard, and a lot of their experience with Airplay (and what features games developers actually want and need) helped shape OpenKODE, so the two are actually very similar in terms of functionality, so the OpenKODE layer is very thin. This means that Ideaworks3D's OpenKODE implementation can be expected to behave identically across different underlying operating systems, while other implementations will probably have variations in minor (but critical if writing a high-end game) details like exactly how the heap behaves, what compiler bugs exist, and things like that. Of course, the whole point of OpenKODE is to provide source-level portability across different platforms, but realistically there will be a few quirks when moving between different OpenKODE implementations or between different platforms on OpenKODE implementations that try less hard to hide platform differences.

Airplay was designed by game developers, so of course it runs as close as possible to the hardware, letting you optimize for the exact platform you're running on and use all its capabilities, such as graphics hardware or floating point hardware. But if you want to write a single build that runs anywhere, you are free to do so.

If anyone doesn't like OpenKODE, Airplay also provides a more traditional set of POSIX libraries, complete with STL etc, with the added advantage that these will behave in exactly the same way across different platforms.''

Symbian continues to provide support for industry standards as part of its core offering. Symbian OS v9.5 will be OpenKODE 1.0 compliant because it will include a reference implementation of OpenGL ES 1.1. Prior to Symbian OS v9.5, only OpenGL ES 1.0 has been supplied, although OpenGL ES 1.1 is available as a plug-in to the S60 3rd Edition.

Symbian OS v9.5 will also offer an implementation of OpenVG 1.01. SVGs are already supported by Symbian OS v9, it's just that the implementation isn't fully OpenVG compliant. Offering full OpenVG compliance in Symbian OS v9.5 is a natural consequence of the platform's continuing evolution.

7.4 OpenVG

The OpenVG standard addresses the growing need in industry for high-quality, scalable, hardware accelerated 2D vector graphics. It mainly targets devices with small screens and low power budgets like mobile phones.

With a drawing model very similar to other vector graphics technologies, OpenVG can provide acceleration for Flash, SVG, PDF and Postscript data and allows the creation of fast anti-aliased graphics that can be used to create screensavers, GUIs and, of course, games. Furthermore, since it is part of the OpenKODE suite, it is also royalty free (which is a big woo-hoo!).

As of mid-2007, OpenVG is gaining a fast adoption rate particularly as its syntax is very similar to OpenGL ES, which makes it very easy for developers to learn and use. Some of its core features include blending, images, image filters, viewport clipping, scissoring, and alpha masking – which are familiar concepts to anyone who has worked with OpenGL or OpenGL ES before.

The current version of the specification is 1.0.1 which dictates that any implementation must "*support all drawing features required by a SVG Tiny1.2 renderer.*"[6]

The specification defines extensive requirements for Bezier curves, 2D transformations, painting and, like OpenGL ES, it uses EGL to acquire graphics contexts and surface for rendering.

As I mentioned at the end of the previous section, Symbian OS v9.5 will be OpenVG 1.0.1 compliant as part of its OpenKODE support.

7.5   OpenMAX

The OpenMAX standard defines cross-platform portability for streaming media such as audio and video codecs, to process a high throughput of data in a consistent manner. It targets middleware developers who produce graphics libraries, multimedia codecs, and other functions for image, video, audio, voice and speech.

In the absence of such a standard, efforts to port existing solutions across platforms have generally been very expensive to date because large amounts of the code base have to be optimized in platform-specific assembler (which sounds like lots of fun doesn't it?). So the OpenMAX standard should provide a large cost benefit in the near future.

There are three flavors of the standard:

- OpenMAX IL (integration layer): defines a standard low-level interface for multimedia codecs on mobile and embedded devices. The main aim here is to provide codecs with a system abstraction layer – the implementation remains transparent to the client.

- OpenMAX DL (development layer): defines an API for audio, video and imaging functionality that hardware vendors can implement on

---

[6] The OpenVG 1.0.1 Specification can be found at ***www.khronos.org/files/openvg_1_0_1.pdf***. A reference implementation for Microsoft Windows is available at ***www.khronos.org/openvg***.

their processors and DSPs. Codec vendors can use these as accelerated building blocks for their own codec implementations. The current version is 1.0.1.

- OpenMAX AL (application layer): defines an application-level multimedia playback and recording API that provides for device-independent cross-platform access to a device's audio, video and imaging capabilities. At the time of writing, this specification has just been released in its provisional form and is expected to be finalized by 2008.

OpenMAX is not currently part of OpenKODE 1.0, but will instead form part of OpenKODE 1.1 to be finalized in 2008. OpenMAX has already gained acceptance and is already supported in popular hardware such as the Sony PlayStation 3 console.

Symbian OS v9.5 will include audio and video codec interfaces that are compliant with OpenMAX IL 1.0 and Symbian plans to support OpenMAX AL in the future as well. This will provide a stepping stone towards Symbian OS becoming a fully compliant OpenKODE 1.1 mobile operating system.

7.6 OpenSL ES

The OpenSL ES (Open Sound Library for Embedded Systems) is currently only at a provisional specification stage. The aim is to define an application-level enhanced audio API that includes features such as advanced MIDI, 3D sound and audio effects. It will be part of OpenKODE 1.1 and I would hope that by this stage of the book, its relevance to mobile game development should be clear to you!

As Aleks has already covered this subject in detail in Chapter 4, I'll say no more about it here other than to note that there is necessarily a degree of overlap between OpenSL ES and OpenMAX AL, specifically in regards to audio playback, audio recording and basic MIDI. This is because OpenMAX AL focuses on media capture and rendering (which includes audio) while OpenSL ES focuses on audio functionality for mobile devices.

Symbian is committed to supporting these emerging standards in the future[7] although it's difficult to say when, since the specs aren't finalized yet. This is kind of like trying to design a system when your client keeps

---

[7] "Symbian plans multimedia standards and bearer mobility" in *Computer Business Review Online*: **www.cbronline.com/article_news.asp?guid=193EEDF0-B8DE-4EF7-91 DB- 5ABA6027558F**.

changing their mind – you know you can provide what they want as soon as they decide just what that is!

7.7   EGL

The Native Platform Graphics Interface (known as EGL) is an interface between rendering APIs such as OpenGL ES or OpenVG and the underlying native platform windowing system. The latest version is EGL 1.3, although currently Symbian OS only has support for EGL 1.1.

EGL handles context management, surface and buffer binding, and rendering synchronization, as well as mixed rendering where OpenGL ES and OpenVG can share a drawing surface. The standard defines three types of rendering surfaces – windows, pixmaps, and pixel buffers (pbuffers). Both windows and pixmaps are associated with native resources, whereas pbuffers are off-screen rendering surfaces used by EGL itself; rendering to pbuffers via the native windowing system may not even be supported.

Using EGL with OpenGL ES in Symbian OS C++ greatly simplifies and standardizes the code needed to obtain a rendering context as compared to achieving the same using `CFbsBitmap`. Setting up EGL for use with, say, Open GL ES generally requires the following steps:

1. Get the main native display using `eglGetDisplay()`.

2. Initialize EGL by calling `eglInitialize()`.

3. Specify the properties of the configuration you would prefer. A configuration includes attributes such as buffer size, colour depth and the existence (or not) of ancillary buffers such as the stencil buffer.

4. Use the `eglGetConfigs()` and `eglChooseConfig()` methods in sequence to find the best match based on the device characteristics.

5. Create a window surface with `eglCreateWindowSurface()` which is where frame-buffer contents will be blitted across to when `eglSwapBuffers()` is called during each iteration of the game loop.

6. Create an EGL graphics context using the `eglCreateContext()` method and the configuration found at step 4. The context stores state information used by the OpenGL ES graphics pipeline and any state changes made by it will be updated in the context object. It will be initialized to default OpenGL ES values.

7. Bind the context, surface and display together by calling `eglMake-Current()`. This tells OpenGL ES which context to use since you can actually have multiple context within the same application.

The setup process described above would normally be performed in the second-phase constructor (the `ConstructL()` method) of a `CBase`-derived class, with the display, context and surface objects being owned members of the class.

It is also vital to free up all EGL resources (as in any resource-constrained environment) and the obvious place to do this is in the class destructor. This normally consists of four steps:

1. Set the current context to nothing with `eglMakeCurrent()`.

2. For each context that was created, call `eglDestroyContext()`.

3. Free up the EGL surface with `eglDestroySurface()`.

4. Terminate the display by calling `eglTerminate()`.

By providing this abstraction layer, the EGL standard promotes standardization across devices and will greatly ease game ports across the wide range of windowing systems available in the mobile device market. It plays a key role in the OpenKODE standard by acting as the communications layer for interactions between sub-systems such as OpenVG and OpenGL ES, the latter of which is the subject of the next section.

7.8   OpenGL ES

As a concept, OpenGL ES ranks right up there with the greatest of all human achievements and no – I'm not exaggerating. If you've actually been lucky enough to witness just how good hardware accelerated 3D graphics can look on a mobile phone, you'll know what I mean. It not only stands on its own as a mature technology, but it also provides support to other higher level libraries, such as the Mobile 3D Graphics API for Java ME.

Symbian has provided a software based implementation of the OpenGL ES 1.0 standard since Symbian OS v8.0a, and both S60 3rd Edition and UIQ 3 SDKs provide plug-ins to upgrade it to OpenGL ES 1.1.[8] In fact, the S60 3rd Edition FP1 SDK includes the OpenGL ES 1.1 plug-in by default.

Currently, Symbian smartphones with support for hardware accelerated 3D graphics include the N82, N93 and N95 from Nokia, the P990, M600 and W950i from Sony Ericsson and the MOTORIZR Z8 from Motorola.

---

[8] You can find the plug-in for S60 3rd Edition development at: ***www.forum.nokia.com/ info/sw.nokia.com/id/36331d44-414a-4b82-8b20-85f1183e7029/OpenGL_ES_1_1_Plug_ in.html*** or by searching for 'OpenGL ES 1.1 Plug-in' from the main page of the Forum Nokia web site at ***www.forum.nokia.com***.

The OpenGL ES SDK for UIQ 3 is available from ***developer.uiq.com/devtools_uiqsdk. html***.

Clearly, this opens the way to a level of quality in game graphics that has not been seen on mobile phones before. The Nokia devices all have the PowerVR MBX graphics processor from Imagination Technologies, which you can read about at *www.imgtec.com/PowerVR/Products/Graphics/MBXLite/index.asp*. Sony Ericsson devices contain the 'Lite' version of the same processor.

OpenGL is a software interface to graphics hardware that has been a 2D and 3D graphics standard on PCs for well over a decade. Although it is the graphics world's workhorse, it wasn't designed to run on embedded devices such as set-top boxes, PDAs, and mobile phones with their relatively limited processor speeds, memory budgets and low power constraints. Also, most embedded devices don't support floating point operations in hardware because FPUs add to manufacturing costs. While most emulate floating point operations in software, this is way too slow for the speeds needed for real-time 3D graphics rendering in a high-end game. In the absence of an FPU, the only way to go is to use a fixed point representation for decimal numbers, which we'll have a brief look at in the next section.

So in 2002, the Khronos Group looked at the OpenGL 1.3 standard and created the OpenGL ES 1.0 specification. To address the wide variance in embedded device capabilities, three profiles were defined – the Common profile, the Common-Lite profile and the Safety Critical (OpenGL ES-SC). The last is intended for use in automotive and avionics displays and uses a minimal implementation, so that the safety certification process is as easy as possible.

Both the Common and Common-Lite profiles are subsets of OpenGL. They introduce support for S15.16 fixed point decimal numbers which can be used for vertex attributes, command parameters, 3D matrix ops etc. The main difference between the two profiles is that the Common profile also supports normal floating point decimal numbers, whereas the Common-Lite only allows the use of fixed point numbers.

When version 1.0 was defined, the devices around were fairly limited, and so they had to be pretty strict about what could stay and what had to go. It was envisioned (correctly) that later versions could include more advanced functionality to take advantage of better hardware as it became available. In the meantime, many redundant or expensive APIs were removed or made optional, data types (doubles) were removed, and new versions of existing functions were added to allow the use of smaller data types like unsigned bytes.

Other features that were removed include support for display lists and `glBegin()`/`glEnd()` since all vertex data in OpenGL ES is specified using vertex arrays exclusively. You also can't specify user-defined clip planes and most of the imaging support has been removed as well. Texturing is also limited – you can't control the level of detail (LOD) in mipmap generation, cube maps are out and so are 1D and 3D textures. For a more detailed summary of the differences between OpenGL and OpenGL ES, it's worth reading the difference specification available on the Khronos website at ***www.khronos.org/registry/gles/specs/1.1/es_cm_spec_1.1.10.pdf.***

OpenGL ES 1.1 is all the rage at the moment. It is based on OpenGL 1.5, which introduced a wide range of new functionality and, while obviously not all of it could be included, some significant features found their way into OpenGL ES 1.1:

- ability to perform most multi-texturing operations (because implementations must now support at least two texture units)

- automatic mipmap generation

- required support for at least one clip plane

- ability to query most dynamic states

- support for vertex buffer objects (VBOs).

This last feature is particularly valuable for embedded devices. VBOs allow you to store frequently used geometry in video memory which means it doesn't have to travel across the bus every time you want to use it. When you put data in a buffer you also specify how you intend to use the data (provided once or provided repeatedly) which allows OpenGL/OpenGL ES to optimize for performance.

Now even though times are changing and we're starting to see some of the first devices on the market with dedicated FPUs, it's still a good idea to avoid using floating point values altogether. At this stage, there's just not enough market penetration and it's usually unnecessary, as fixed point arithmetic will generally provide sufficient accuracy for rendering and simulation calculations.

So it's 2007/2008. OpenGL ES 1.1 is widely used, and there's support for development on the latest Symbian smartphones using the most recent SDKs. We're seeing the first hardware accelerated 3D graphics handsets appearing on the market using the PowerVR GPU and it's all new and exciting. There's only one thing missing – shaders.

Vertex and fragment shaders are simply small programs that replace sections of the traditional fixed function pipeline. Apart from the fact that using shaders allows a wide range of operations and effects that are just not possible with fixed functionality, the important thing is that when you use shaders you have far more control over what is done at key steps. So this allows you as the programmer to optimize rendering and reduce power consumption, since you get to choose which operations are applied and which aren't.

OpenGL 2.0 introduced shaders and frame buffer objects, which allow you to render to texture – so you can draw a scene and then use that image as a texture for some other world object, if that's what you want to do. High level shaders are written using the OpenGL Shading Language (OpenGL SL) which looks a lot like C but has specific constructs for graphics programming.

For embedded systems, the equivalent version is OpenGL ES 2.0 (**www.khronos.org/opengles/2_X**); clearly, the optimizations outlined above are going to be even more valuable when you're working with mobile phones. The shading language also has a cut-down version called the OpenGL ES SL (no surprises there). The spec for OpenGL ES 2.0 was only finalized in March 2007, so it's probably going to be a while before we see support in phones – although having said that, Imagination Technologies' PowerVR SGX GPU series is already OpenGL ES 2.0 compliant, and they have already released an SDK for PC emulation.[9] Given that standards are always evolving to meet new challenges, it shouldn't be long until Symbian smartphones support this amazing technology; Symbian has penciled in Symbian OS v9.5 as the one that delivers a reference implementation of OpenGL ES 2.0. I just hope it's not too long, because when we get hardware accelerated 3D graphics using shaders, it's going to make a lot of FlashLite games look '*so* 2007'!

7.9   Get Your Fix

While it's all very well to go on about how good all these standards are, I wanted to finish off on a slightly different note. This section doesn't talk about any standard in particular, but I thought it would be appropriate to present a quick discussion of fixed point numbers, since they play such an important part in OpenGL ES development.

As we've already said, many manufacturers don't put FPUs in handsets due to increased production costs and power constraints. Conversely, sometimes FP calculations are emulated in software instead – but these are too slow to be useful in games; even if they weren't, their power consumption is also prohibitive – a rather interesting Catch-22.

So the solution is fixed point mathematics – using integer arithmetic to yield decimal results. But like everything in life, it comes with a price; what you gain in speed, you lose in precision and in representational range. After all, an integer is *not* a decimal, so clearly an approximate representation needs to be employed.

A decimal number consist of two parts – the integral part to the left of the decimal point, and the fractional part to the right. Now, in a fixed point representation, the location of the decimal point is fixed (wow – how insightful). So if you have a number made up of N bits

---

[9] The SDK can be found at **www.imgtec.com/PowerVR/insider/toolsSDKs/ KhronosOpenGLES2xSGX**.

and you use M bits for the integral part, you have N-M bits left over to represent the fractional part.

You'll also want to be able to represent negative numbers so the integral portion is in fact a 2's complement number with the most significant bit used as the sign bit. If we use a 32-bit integer and reserve the bottom 16 bits for the fractional part, we say that we have either a 16.16 fixed point number or an S15.16 fixed point number (where the 'S' indicates the sign bit).

Now let's see what this looks like. In fixed point arithmetic, all the digits to the right of the 'decimal' point are multiplied by negative exponents of the base. Sounds scary, so check this out this representation of π in base 10:

```
(3.14159)₁₀ = 3 * 10⁰ + 1 * 10⁻¹ + 4 * 10⁻² + 1 * 10⁻³ + 5 * 10⁻⁴ + 9 * 10⁻⁵
            = 3 +        0.1 +       0.04 +      0.001 +    0.0005 +    0.00009
```

Makes sense doesn't it? Now let's convert another example expressed using base 2:

```
(110.01000)₂ = 1 * 2² + 1 * 2¹ + 0 * 2⁰ + 0 * 2⁻¹ + 1 * 2⁻² + 0 * 2⁻³ +
               * 2⁻⁴ +  0 * 2⁻⁵
             = 4 +      2 +      0 +       0 +        1/4 +      0 + 0 + 0
             = 6.25
```

But wait! This is a binary number, so we can do bit ops, like shifting, on it. So let's get rid of the fractional part by shifting left by 5. We now have 11001000 which is equal to 200. It doesn't matter whether we use binary or decimal, the result is the same, so we can now say that $6.25 = 200 \gg 5$. And, whacko, we've represented a decimal number by using an integer with a bit shift.

The final piece of cleverness is that if you're using a fixed format such as 16.16, which is what OpenGL ES uses, you know that you'll always be shifting right or left by exactly 16 bits. This is the same as multiplying or dividing by $2^{16} = 65536$ respectively, and should explain the meaning of the following four macros found in utils.h in the S60 SDK OpenGL ES examples:

```
#define INT_2_FIXED(__a) (((GLfixed)(__a))<<16)
#define FIXED_2_INT(__a) ((GLint)((__a)>>16))

#define FLOAT_2_FIXED(__a) ((GLfixed)(65536.0f*(__a)))
#define FIXED_2_FLOAT(__a) (((GLfloat)(__a))* (1/65536.0f))
```

You should be able to see for yourself that the normal arithmetical operations like addition and multiplication are actually performed as integer operations, which are obviously a lot faster.

However, a word of caution is in order here: in non-trivial calculations, the danger lies in temporary results occasionally being inexpressible in the fixed point format being used – this can happen when there are not enough digits to express the number (overflow), or when the number is too small to be represented (underflow).

Now that you know a bit about fixed point arithmetic, it's worth remembering that the Symbian OS build tools will build user-side code as Thumb unless you tell them not to by using the BUILD_AS_ARM specifier in your bld.inf file, or the ALWAYS_BUILD_AS_ARM keyword in the project MMP.

A key benefit of using the ARM instruction set when using fixed point numbers is that you get 'free' support for bit shifting using the ASR (arithmetic shift right) and the LSL and LSR (logical shift left/right) instructions. This is because while these instructions are also present in the Thumb instruction set, under ARM, the bit-shift ops can be piggy-backed into another instruction such as MOV, ADD, SUB and so on; therefore, the shift doesn't have any impact on the instructions' execution time. Clearly, this can translate into a huge performance gain when you're heavily using fixed point arithmetic for physics simulations and 3D rendering using libraries such as Mascot or OpenGL ES.

Hopefully, this brief introduction will help you use the GLfixed data type and APIs in OpenGL ES. For a thorough treatment of fixed point arithmetic, refer to [2] and [3] in the Further Reading section below.

7.10   Enough Already!

We've covered a lot of ground in this chapter, and it doesn't have code in it to break it up so, if you got this far, well done! You may have also noticed that, while the chapter title is 'C/C++ Standards Support for Games Developers on Symbian OS' there was nothing on support for standard C++ libraries like the STL. At the time of writing, there is no official support for standard C++ on Symbian OS, although there are a number of unofficial versions.

As regards POSIX, it's also unlikely that Symbian OS is going to introduce support for select, fork, exec or signal in the short or medium term, without a significant re-design of some of the core operating system parameters. Regardless, there is nearly always a way around the problem, outlined in detail by the guys who built P.I.P.S. in the *Using P.I.P.S.* booklet from Symbian Press, found at ***developer.symbian.com/main/learning/press/books/pdf/P.I.P.S..pdf***.

Symbian OS has evolved over time to meet a wide variety of industry standards while maintaining its position as the premier mobile phone operating system, which is no mean feat. This support should make porting games to Symbian OS a fairly easy task both now and in the future. But if you're considering the use of a middleware solution to ease the effort of porting between different platforms, you should turn to the Appendix on Ideaworks3D's Airplay SDK, found at the end of this book.

7.11 Further Reading

[1] *Core Techniques and Algorithms in Game Programming*, D. Sanchez-Crespo Dalmau. New Riders, 2003.

[2] *Mobile 3D Graphics*, A. Malizia. Springer-Verlag, 2006.

[3] *OpenGL ES Game Development*, D. Astle and D. Durnil. Premier Press, 2004.

8

The N-Gage Platform

Peter Lykke Nielsen and Roland Geisler (Nokia)
Jo Stichbury

8.1 A Brief History of N-Gage

Mention the N-Gage 'game deck' mobile device by Nokia to anyone who has worked in mobile games, and you'll find they have an opinion. The N-Gage devices continue to polarize opinions – people either love them or hate them!

As a mobile device manufacturer, Nokia has always been committed to mobile games. It was the first manufacturer to put a game, **Snake**, onto a mobile device, back in 1997. Nokia found that the game caught on and quickly became popular. Feedback was so good that Nokia started to include more games as part of their ongoing strategy, including embedded games, such as **Space Impact** and **Pairs**, in their devices. Nokia owners could register to become part of Club Nokia and post their game high scores – an early example of social mobile gameplay.

8.1.1 N-Gage Game Decks

From the start, Nokia was quick to develop a comprehensive strategy to bring rich content games to the mobile platform. The strategy included device hardware and software technology components, plus marketing and distribution services. The combination was given a separate brand: N-Gage. Prior to N-Gage, the game industry for phone-enabled mobile devices consisted mostly of small development studios that developed Java games. Nokia recognized the long-term opportunity to enrich the user experience by bringing a mainstream video game experience to their mobile phones.

Nokia's S60 UI platform on Symbian OS was released in 2002 (at that time, it was known as Series 60). The N-Gage game deck was one of the first devices to be released on the platform,[1] and hit the streets in October 2003. As you'll see from Figure 8.1, it was designed to be attractive to people who wanted to play high-quality games on their mobile phone, having controller keys optimized for playing games. Besides the ability to play games, the original N-Gage game deck had features typical of the Series 60 platform at the time, including messaging and PDA functionality (contacts, calendar, to do list and so on), as well as the ability to make voice and data calls. It also played MP3s and had an FM radio player, Bluetooth and USB local area connectivity, and GPRS network connectivity.

Figure 8.1   The original N-Gage game deck

The original N-Gage game deck had controls familiar to gamers, arranged in a similar configuration as on a typical handheld console, making it attractive for gameplay. The game deck had some necessary design constraints that may have put off those with only a passing interest in games. One example was the ergonomics – the game deck was designed so the user held it sideways, speaking into the narrow edge, rather than flat against the face. Another constraint at the time was that hot-swapping between external multimedia cards was not supported by the platform. Games were delivered on multimedia cards, and to swap between games, the user had to remove the back of the game deck and take out the battery, forcing the device to reboot.

Nokia responded to feedback about the N-Gage hardware, and a year later released a second game deck called the N-Gage QD, shown in Figure 8.2. The QD was cheaper, could be held more naturally when using the phone, had a better battery life and, most importantly, allowed hot-swapping of game cards. The USB port, FM tuner, and MP3 player were, however, removed from the QD to allow it to be smaller in size. Sales of this new game deck were successful and it became very popular in some regions; however, it is no longer manufactured.

---

[1] The N-Gage game deck mobile devices were based on the 1st Edition of the Series 60 platform.

Figure 8.2 The N-Gage QD game deck

Although the sleeker N-Gage QD was a greater success than the original N-Gage game deck, Nokia found itself head-to-head in the portable/mobile game market with Sony's PlayStation Portable and the Nintendo DS. Although very different in functionality, and intended to be rather more functional than those portable game consoles, the specialized look of the N-Gage game deck devices meant they were competing for the same market share, against newer hardware coming from the behemoths of the gaming industry. Nokia found itself in an aggressive market sector, competing against the two most successful players in the market – and it was time for a re-think. Instead of making a few specialized gaming phone products to run the high-quality mobile games they were creating, wouldn't it make more sense to allow a greater range of Nokia's S60 phones able to run those games? To make devices that could also play great mobile games, rather than make mobile game players that were also phones?

Nokia was selling millions of S60 smartphones, and it made sense to make high-quality games available to the people buying them. That's where the new N-Gage platform comes in, as we'll discuss in most of the rest of this chapter. But first, let's talk a little about the games that were available on N-Gage game decks.

8.1.2 Games for the N-Gage Game Decks

Over the lifetime of the N-Gage game decks, over 30 N-Gage games were released. Some of these are cited as great examples of mobile games, including *Pathway to Glory* and *High Seize*, both of which took advantage of the multiplayer functionality available from the N-Gage Arena, discussed later in this section and in more technical detail in Chapter 5. Games were released on multimedia cards, which could be bought from shops and outlets selling boxed game products. They were not available to download over the Internet – the size of the games ruled out downloading over the air – until very late in the life cycle of the N-Gage game decks.[2]

Inevitably, the first games released were affected by piracy. Within a month of releasing the N-Gage game deck, there were reports of games being 'cracked' (which meant they could be downloaded and installed without purchase). Nokia moved quickly to implement a new security system to protect the intellectual property of their games and prevent games becoming available through illegal channels. However, the loss of profit, due to infringement and distribution of cracked games, was a dent to the image of the N-Gage brand.

Nokia wanted the games available for the N-Gage game deck to be of a very high quality. At the time the N-Gage game deck was conceived, games for mobile phones and portable game consoles often had poor graphics and audio, and were typically 'casual' games, such as card games and well-known titles like **Tetris**. Game developers were mostly using Java ME, where the games were constrained by the performance characteristics of the Java platform, and by the fact that many of the phones they were targeting had very limited main memory and slower CPUs.

The capabilities of the N-Gage game decks could enable game developers to create more imaginative titles, use better graphics, sound effects, and music; they could also cater for people who wanted a game that provided hours of interest, challenging gameplay, and high-quality presentation. Nokia needed good games, and to ensure this, the company created a business unit to publish mobile games themselves. Chapter 1 discusses the role of publishers in the game industry; in effect, the N-Gage game publishing team commissioned game titles from external game developers, managed the end-to-end game creation process using in-house game producers, play-tested and certified the games, and then marketed these games – and the N-Gage Arena community – to the game-playing public.

The first generation of games for the N-Gage game deck was outsourced to well known developers such as EA, Vivendi, and THQ. Nokia developed strategic relationships with these industry leaders to deliver their titles and game licenses, such as **FIFA Soccer**, **Splinter Cell**, **Lara Croft** and **Tiger Woods Golf**, for the N-Gage game decks.

The N-Gage game publishing team comprised a number of experienced industry professionals, and was well-placed to observe the evolution of the mobile game market. The N-Gage game decks may no longer be available, but their legacy is a good one. The creation of a specialist team for innovative mobile game publishing allowed Nokia to build a

---

[2] For those of you who still have N-Gage game decks – some of the most popular games can still be purchased online at **www.softwaremarket.nokia.com**, with free trials available before you buy.

wealth of experience of the mobile game market. The game publishing team was well-placed to define the evolution of the N-Gage game deck into a platform, and its experience has undoubtedly given Nokia the competitive edge over other handset manufacturers, and a lead position in the mobile game market as a whole.

8.1.3   N-Gage Game Developers

As a publisher, Nokia worked with a number of game developer companies worldwide to create games for the N-Gage game decks. These developers created native games written in C and C++ using an SDK provided specifically for them. The N-Gage SDK was an extension of the generic Series 60 1st Edition SDK available to the public at the time. It was created in Bochum, Germany by a team that was co-located with the N-Gage device creation team. The N-Gage SDK provided a number of additional libraries to access functionality that the standard Series 60 SDK did not provide (such as content protection for N-Gage games, game playing protocols over a Bluetooth connection, and utilities for detecting system events relevant to games, such as charger disconnection events and low battery and network signal notifications).

One of the other features of the N-Gage SDK was the ability to access the APIs necessary to add networked multiplayer functionality to a game, taking advantage of the connectivity of the N-Gage game deck and differentiating it from other mobile game consoles. The technology behind the N-Gage Arena is discussed in more technical detail in Chapter 5. In brief, it provides multiplayer gameplay over a GPRS network including a backend system that provides hosting, data storage and user account management. The N-Gage Arena SDK provided APIs for developers to use that presented a standard user interface for handling network logins, lobby and game room creation and matchmaking. Developers were able to focus their efforts on game development instead of creating a networking game library. For example, using the APIs provided by the N-Gage SDK, a game developer could either add the ability for a game to be played simultaneously by two or more players (for example, a turn-based action game or a shadow racing game). The N-Gage Arena APIs also provided the functionality to upload game high scores and then send messages to challenge friends to beat them, along with a range of other connected gaming features.

The technology behind by the N-Gage Arena was originally part of Sega.com's SEGA Network Application Package for mobile multi-player gaming. Nokia acquired Sega.com Inc. (a subsidiary of SEGA) in early 2003 to enable them to offer networked multi-play and a virtual community to N-Gage gamers immediately at launch.

The N-Gage SDK was provided free of charge to developers contracted to write N-Gage applications, and was supported by a team dedicated to helping with technical issues, as well as a number of custom tools and documentation. The N-Gage SDK was the first Nokia SDK that provided support for on-target debugging, as well as other tools to help the game developers debug their code and test it prior to game certification – which is something we'll discuss next.

Although it was, and still is, possible to write games for the N-Gage game decks,[3] games are not official N-Gage titles unless they are published by Nokia and meet their industry-standard game requirements. The N-Gage standard game requirements (SGRs) were defined based on the behavior and quality levels expected of a professional-quality game and, besides play testing, each title was tested against the SGRs. The testing was called certification, and was key to ensuring that all the games released for Nokia were of a consistent standard.

8.1.4 Summary

Nokia can be said by some to have been, at best, before their time, and, at worst, to have taken the wrong direction with the N-Gage game deck hardware. However, it is generally agreed that the commercial games delivered for N-Gage by Nokia as a game publisher were of exceptionally high quality, and made creative use of the phone hardware, networked multi-play, and phone capabilities available at that time. The games built up a loyal fan base, which is supported to this day through the N-Gage website (**www.n-gage.com**) and the N-Gage Arena community.

Having discussed the history, to put it all into context, it is now time to move on to describe the present day. Instead of producing another gamer-specific phone, Nokia has decided to concentrate on providing smartphones that appeal to everyone, and to help developers create and deliver great games to play on them. Enter the N-Gage platform!

8.2 N-Gage Platform: The Next Generation of Mobile Gaming

The next generation of N-Gage, the N-Gage platform, was first announced at the E3 Show in Los Angeles back in early summer of 2006, and is due for release in early 2008.

Rather than delivering a single handset, Nokia makes it possible to play N-Gage games on a range of S60 smartphones, allowing the user

---

[3] It is still possible to develop games (or other applications) that run on the N-Gage and N-Gage QD game decks. To do this, you need to use the S60 First Edition FP1 SDK available from **forum.nokia.com**. The N-Gage SDK is not available for public use.

to choose between different styles, form factors and prices. This enables Nokia to direct its high-quality N-Gage games to a much larger and more diverse user base.

N-Gage compatible smartphones must meet a minimum set of terminal requirements to give game developers a common set of hardware and functionality when creating their games. All phones are expected to have a minimum allocatable heap size of 10 MB, a minimum storage capacity (on an external memory card or built-in memory) of 64 MB, and support for standards such as Bluetooth 1.2 and OMA DRM 2.0. Some phones may be game optimized with dedicated controller keys and landscape-orientation gaming (where the screen is sideways, with the longest edge horizontal as is typical for portable game consoles) as well as portrait-orientation gaming.

At the time of writing, Nokia has just announced the first range of S60 3rd Edition smartphones that will be N-Gage compatible – these are shown in Figure 8.3 and include the Nokia N81, N81 8 GB, N82, N95, N95 8 GB, N93, N93i and N73. More information about the phones can be found at **www.n-gage.com/get_ngage/devices.html.** As new S60 smartphones are created, they are assessed against the N-Gage terminal requirements to determine whether they have the minimum capabilities needed to support N-Gage games. It is to be expected that a large number of Nokia S60 smartphones will eventually be N-Gage compatible, with a correspondingly-sized audience of potential gamers.

Figure 8.3   N-Gage compatible S60 smartphones announced in August 2007

8.2.1   What the N-Gage Platform Offers Users

Nokia provides an easy-to-use and reliable service for users to acquire the games through the N-Gage application. By allowing an easy way to discover, try, buy, play, share and manage mobile games, the N-Gage platform sets out to be an exceptional mobile gaming experience. The aim is to remove the woes of the past with regards to finding a game compatible with a particular phone, working out how to purchase, download and install it, some of which we described in Chapter 1. Nokia is investing heavily in the supporting technologies and back-end infrastructure required for this service to function.

The overall platform experience will be managed from a client application called the N-Gage application. This application is guaranteed to work on every phone supported by the N-Gage platform and can be downloaded from the web if it is not already pre-installed on the phone when it is purchased. The N-Gage application is described in more detail in section 8.4.

At launch, Nokia is expected to have numerous games available for N-Gage, with a variety of genres and licensed intellectual property, including **Pro Series Golf, Snakes Subsonic, Mile High Pinball, Hooked On: Creatures of the Deep** and **System Rush: Evolution**. More information about the games available can be found on the Showroom page of the N-Gage website at **www.n-gage.com/showroom.html**.

8.2.2   What the N-Gage Platform Offers Developers

The issues caused by differences between phone handsets (device fragmentation) are constantly being addressed by handset manufacturers, like Nokia, that appreciate the need for a common set of features across a range of phones, and use the S60 platform to provide it. However, even with this platform, developers still have to put in effort to create and support a native game across multiple devices – each differing in the amount of memory available, screen resolution and display size, input capabilities, and other hardware characteristics. A great deal of the development budget and time is spent on managing device differences, time that should really go towards tailoring the games for an optimum playing experience on the mobile form factor.

Nokia decided that, to make it easier for professional game developers to reach as many consumers as possible on the S60 platform, the first requirement in creating the N-Gage development environment was simple. The developer should be able to create one SKU (stock keeping unit), that is, one deliverable, that runs on all devices.

The N-Gage SDK addresses the issue of portability between different S60 smartphones by providing a set of libraries that presents a consistent interface for the games to establish hardware capabilities and manage the differences between them. The SDK provides an abstraction layer between an N-Gage game, the services running on the device, and the phone hardware. The game code is insulated from the S60 platform and Symbian OS, and it must use only the APIs provided by the N-Gage SDK.

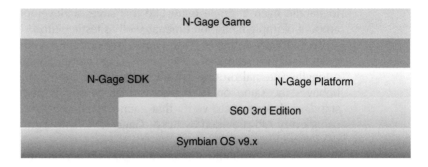

Figure 8.4   The N-Gage SDK abstracts Symbian OS and S60 from game developers

Although this can be limiting in certain respects, the abstraction layer guarantees compatibility across different devices, taking the responsibility away from the game developer.

As Figure 8.4 illustrates, the N-Gage SDK provides a single point of contact for the game in terms of how its code is exposed on a device. As the number of devices supported by the N-Gage platform grows, older games will automatically be supported by these newer devices without the developers being forced to handle any expensive code maintenance and conversion.

A good example of how the N-Gage SDK works as an abstraction layer in practice can be illustrated by considering the issues required to support game playing in landscape orientation on different phones. For example, some phones support portrait orientation only, others support both portrait and landscape; some have two variations of landscape-orientation, while, finally, some allow the user to twist the device into a certain orientation. Moreover, on some devices, certain orientations cover the keypad and prevent numeric input. Handling different orientations using the native Symbian OS APIs was discussed briefly in Chapter 3.

The set of S60 phones that are N-Gage compatible will grow to add more phones and, of those, a large number of additional phones will come to market after a game is shipped. How is the developer to implement the game code? It would require a logic flow that takes into consideration the behavior of each N-Gage phone (current and future) so that the game can adjust the graphics for the screen mode and input keys accordingly. As you expand the number of devices the decision tree for handling all this logic becomes untenable in a single SKU.

When using the N-Gage SDK, a developer does not have to manage all this logic, since it is handled for them at a platform level. The SDK provides the developer with a handful of methods to detect what orientations a phone supports, what the current orientation is, and how to set a new orientation. The underlying alignment of display and keys is handled in the background by the platform. Avoiding the code required to handle even a single issue like this saves a developer a significant amount of implementation time, as well as testing effort.

To guarantee to the N-Gage game development community that their games will port seamlessly between N-Gage compatible phones requires a lengthy internal process within Nokia to ensure that the N-Gage SDK behaves the same on each phone. SDK testing is performed using a standard set of tests to verify that each device behaves correctly when using every API provided by the N-Gage SDK. This is called terminal, or device certification.

The device certification teams also play an important role in verifying the technical solutions to known issues where different behavior exists across devices. For example, a game should include a screensaver for display when the game is idle. At the S60 architecture level, this gets complicated, because the native screensaver functionality varies across different S60 product versions. The N-Gage platform must hide this, and expose a simple interface to the game. The solution must also provide flexibility to incorporate new behaviors in the future without breaking binary compatibility. As should be clear from the above examples, the technical ambition behind the promise of providing an abstraction layer in the SDK is certainly a challenging one.

8.3   The N-Gage SDK

The experience of the N-Gage game publishing team showed Nokia that conversion of a game title from other platforms' hardware was also not straightforward. The mobile phone as a computing device has very different requirements from other platforms. Symbian OS was designed to be fit for purpose and, while the operating system is ideal for mobile telephony, specialist applications such as personal information management and enterprise support, the programming paradigm is quite different from that used to create games for other platforms. This typically resulted in the game developers re-writing a large amount of the code base to convert it for N-Gage devices, making it an expensive option for game ports.

In addition to the abstraction layer described above, the N-Gage SDK also abstracts Symbian OS and provides a POSIX compliant, standard C/C++ layer over Symbian OS. The result is a much more straightforward conversion of game code from other established platforms, because game developers, familiar with other platforms, no longer have to learn Symbian OS C++ idioms, like active objects and descriptors, before they can port their code. As a result, large parts of the game logic can be ported more or less directly to the N-Gage platform, leaving the development team with time and resources to focus on addressing special design issues with the game relevant to the N-Gage platform.

The N-Gage SDK extends the initiative from Symbian and Nokia whereby the P.I.P.S. (a recursive acronym for 'P.I.P.S. is POSIX on Symbian OS') and Open C libraries make standard C libraries available for Symbian OS v9 devices. P.I.P.S. and Open C are described in more detail in Chapter 7. They add standards support for basic operating system functionality, including:

- APIs for string handling, file system access, maths, thread handling, and DLL loading (`libc`, `libm`, `libpthread`, `libdl`)

- some additional open source libraries for cryptography and SSL (`libcrypt`, `libcrypto`, `libssl`) and compression (`libz`)

- an additional general purpose utility library (`libgl`).

Open C does not supply any access to the special capabilities of the S60 hardware, such as camera, vibra, or telephony, nor does it offer interfaces to S60 application data such as calendar, contacts, messaging, or multimedia playback.

8.3.1 RGA APIs

The N-Gage SDK adds a set of real-time graphics and audio (RGA) APIs. These APIs are targeted specifically at games requiring high-quality graphics and multimedia content, and were developed with input from professional game developers working with Nokia. In conjunction with Open C, the RGA APIs enable developers to write games using familiar C/C++, rather than using the idioms of C++ on Symbian OS. At the time of writing, these 'industrial-strength' game APIs are part of the N-Gage SDK only, which means that they are currently only available to official N-Gage developers. However, the RGA libraries are expected to become more widely available to S60 developers in the near future, with access to the APIs that were made part of the general S60 offering, making the libraries available for use by all developers, albeit supported on selected S60 devices only.

The main features of the RGA APIs are described in the following subsections, and illustrated in Figure 8.5.

Device Information

APIs are provided to obtain information about the phone's capabilities and status information.

The Device Capabilities API provides developers access to information from an S60 device that is used to determine if the device is qualified for N-Gage game playing, and to get information on the CPU and whether additional features exist that allow enhanced game functionality, such as floating point support and the amount of RAM available to the game.

Figure 8.5 RGA libraries

The Device Status API is a library which provides interfaces to retrieve the current status of various hardware features, including specific information about the battery, network signal, connected accessories (such as headsets), telephony, alarms, and current profile.

The API can also be used to receive notification of changes to the device status, through a set of callback interfaces for the following:

- changes to the battery level

- connection or disconnection of the main charger

- changes to the network signal

- changes to the GSM cell ID or location area code of the phone

- connection or disconnection of an accessory

- changes to the phone's profile (such as the settings for keypad tones, warning tones, flight mode, ring tones volume, display language)

- notification of an incoming phone call, and its termination

- notification of an alarm.

Hardware Access

These APIs provide controls for the device hardware, such as the lights (screen backlight for the primary and secondary displays, and keypad lights) and the vibra for phone vibration. The APIs can also be used to provide dynamic information about changes to the hardware, such as changes to the vibra setting.

Display

The Display API provides system-independent window handles for all of the phone's displays. These window handles are required to create a back buffer on the specific display. The API provides notification of changes to the orientation of the display – and can also be used to return the display settings and capabilities – and allows adjustment of some of those settings, if the underlying system supports it. Display settings include brightness, contrast and gamma correction values. The capabilities returned by the Display API include the native size and resolution, orientation, and color formats.

Input

The Input API provides information on the states and state changes of different input devices, such as embedded keypad and four-way controller, and external input devices connected to the phone via USB or Bluetooth wireless technology.

The Text Input API is also provided for handling text input on keypads in multi-tapping, predictive text input and numeric mode. The Text Input API supports Western typing as well as the following Chinese text input spellings: stroke, Shuyin, and Pinyin.

Timing

The Timing API provides standard and high-resolution timers. Both periodic and one-shot timers are available:

- Periodic timers: After starting the timer, it runs for the given time. After the time has expired, it restarts automatically. The timer will continue to run until it is stopped or released.

- One shot timers: After the timer is started, it runs until the set time period is exceeded. The timer has to be restarted manually after it expires.

Image Rendering

The Back Buffer API allows drawing to the screen without tearing and flickering effects. It also allows changes to the orientation of the screen content.

The API provides a device-independent double-buffered graphics scheme consisting of a screen buffer and a secondary buffer called the back buffer, which can be accessed and drawn into. When drawing is completed, the contents of the back buffer are moved to the screen buffer and become visible on the screen. Single or chained back buffers are supported.

The Bitmap API is an interface to create and manipulate bitmaps and includes the functionality to:

- perform masked blits from one graphics device to another graphics device, such as, a bitmap or the back buffer
- perform alpha-blended blits from one graphics device to another graphics device
- use self-defined color palettes
- scale and rotate the content of graphics devices
- access and modify pixels.

The Images API allows you to load and save images from files of different color formats, and load images from memory buffers.

Fonts

The Fonts API provides a system-independent interface for drawing text to a graphics device (back buffer or bitmap surface) using system fonts with a specific font face, size, and style. It also provides an interface to create game-specific bitmap fonts and draw text to a graphics device on a particular point. Common features for system fonts and bitmap fonts include:

- drawing text to a graphics device (back buffer or bitmap)
- retrieving the size occupied by the text drawn
- applying text effects like underline and strikethrough to the text
- drawing text in different directions
- drawing text with the user-specified color
- drawing text to the graphics device from the given target point.

The following system font features are available:

- installing and uninstalling system fonts
- getting the installed system font count
- getting information about the currently installed fonts on the system
- converting unit size in twips to pixels.

The supported system font formats are TTF (True Type fonts) and GDR (Symbian OS bitmap format). Bitmap font support includes:

- setting the kerning information between the characters
- setting the transparent color for the text in RGB format
- selecting whether transparency should be used or not
- setting default spacing between the characters.

2D Vector Graphics: SVG-T

The Scalable Vector Graphics – Tiny (SVG-T) API provides interfaces to load static or animated SVG files and to perform graphical operations such as pan, scale, and rotate. Support is available for SVG-T, SVGB (prepared SVG in binary format), SVGZ (Zipped SVG), PNG, and JPEG files. The SVG-T API returns a bitmap object that may be blitted onto the back buffer.

Audio

The High-Level Audio API provides a set of functions for playing, recording, and mixing audio. Typical use cases supported include the following:

- retrieving the capabilities of the audio mixer
- initializing and configuring the mixer
- playing back sound on the mixer and stopping the mixer
- adjusting the mixer playback settings
- adjusting the individual playback settings
- pausing or stopping playback of a sound
- managing sounds in logical units (tracks)
- recording audio data.

An additional Compressed Audio API is available for configurable low-level audio playback and recording. This includes support for playback of MIDI and compressed audio files such as MP3 files, audio stream playback (buffer-based), and audio clips playback (file-based).

Camera and Video Playback

The Camera API supplies interfaces to access the phone's camera devices including functionality to:

- enumerate the available cameras
- query the capabilities of each camera to return information about the orientation of the camera, the possible optical and digital zoom values, as well as the supported image capture sizes

- get and set the configuration of a camera, including the zoom factor, contrast, brightness, flash mode, exposure mode, and white balance adjustment

- use the viewfinder to switch between what the camera sees and what the user would see if the device was a mirror. It is also possible to specify a clipping rectangle

- capture images, and specify the color format and image size of the captured image. It is also possible to specify a clipping rectangle.

A Video Playback API provides support for playing or pausing a video clip, and for configuring the playback image (the playback area size, position and orientation, the current and end position, and the playback volume).

Memory Management

This API allows the flexible use of multiple heaps and implementation of custom memory allocation algorithms for those heaps, according to the developer's own memory usage routines.

Runtime and Application State

The Runtime and Application State APIs are provided to create a framework for a game, to load the other APIs, and to handle asynchronous background tasks and notifications.

The Runtime API loads the libraries providing the APIs described above, and is used to instantiate each as it is required. It also initializes the system, yields to the operating system to handle background events, and monitors the time elapsed since the game loop started running.

The Application State API notifies the game of focus loss and focus gained events, and of events occurring when the operating system requests the game to shut down.

The APIs described above constitute the RGA libraries and are correct at the time of writing. The libraries may be modified slightly before they become more widely available, and you should consult reference material and example code provided with them, on their general release.

8.3.2   N-Gage Platform APIs

Besides the APIs discussed, the N-Gage SDK also supplies a number of APIs specifically for use in N-Gage games. For example, APIs are available for games to store game data, to provide game system services for multiplayer game playing over a Bluetooth connection and for access to the N-Gage Arena (including APIs to access the Arena Framework and the Matchmaking and Ranking, Asset Management, and Chat APIs).

As an example, the Game System Services API allows games to add the enablers for interaction with the N-Gage application, which is described in section 8.4. They provide an easy way to display in-game notifications, to perform the start-up check, and to switch to the N-Gage application while playing the game. (In-game notifications are messages that are displayed during the game. For example, messages could be, "John is online" or "Tom says Hello." These messages will be triggered by the N-Gage application.)

N-Gage specific APIs such as the Game System Services API are of use only to those working with the official N-Gage SDK, and will not be described further here. Information about the N-Gage Arena can be found in Chapter 5.

8.3.3   N-Gage Tools

The N-Gage SDK provides a number of additional tools, which are described in the following subsections.

Game Packaging

Digital Rights Management (OMA 2.0 DRM) content protection is used to prevent game piracy. A flexible packaging tool is provided to create a proprietary format N-Gage install file which contains various game metadata, activation codes and retailer information. A Digital Rights Management (DRM) encryption tool is provided, as is a set of documentation and test data for use when preparing to package the game. DRM keys, metadata files, activation codes, and game-specific file access components are generated for each game, on request, to ensure content protection is unique to each game published.

Game Wizard

Another tool included in the N-Gage SDK is a game wizard which can be used to set up the game loop. It provides boilerplate code for initializing the Runtime, Back Buffer and Input APIs. The game wizard is implemented as a plug-in to the Carbide.c++ development environment (Carbide.c++ is recommended as the standard development environment and is a Nokia product specifically for use in C++ development for S60).

MemChecker

The N-Gage MemChecker runs on the phone and is used to monitor memory usage while a game is running, for example, to facilitate testing the game under low memory conditions.

Features include:

- detailed output of memory consumption
- data on the memory, and maximum memory, used by the game
- ability to set limits for the heap size of the game to test under low memory
- simulation of memory failures
- memory leak monitoring.

Trace Viewer

The N-Gage Trace Viewer is used to observe the execution of an application at near-to-run-time conditions, having minimum performance impact on the traced application. Features included are:

- tracing in the Windows emulator
- tracing on the phone, including tracing of release, rather than debug, builds
- transmitting collected trace data to a host, regardless of the current state of the traced application (a crash or lockup of the application does not have any side effects on the tracing process)
- multi-process and multi-target tracing.

8.4   The N-Gage Application

We mentioned briefly in section 8.2 that the N-Gage application and N-Gage games run on a number of different phones, so the user is not forced to buy a single, specific device in order to play N-Gage games. The N-Gage application will be pre-installed onto most N-Gage compatible phones, but can also be downloaded from *www.n-gage.com* for owners of some of the older Nseries phones (e.g., N73).

The N-Gage application is the entry point to the N-Gage experience, collecting all the games into one place on the phone, so it is easy to find and play them once they are installed. A key feature of the N-Gage application is the online game showroom, where all the available N-Gage games can be found and downloaded. All of the games support the 'try and buy' method, which means that a player can try the game in a demo version before deciding to buy it. The opportunity to play before purchasing is a big part of the N-Gage experience, and games are designed to allow them to demo their main features free of charge.

The N-Gage application consists of five modules: Home, Showroom, My Games, My Profile, and My Friends.

8.4.1 Home

The Home module, shown in Figure 8.6, is the entry point into the N-Gage experience. From here, players can quickly start or resume a game, track their progress and see a summary of their game profile, choose friends to chat with, check out new games, and access private messages from friends.

Figure 8.6 The Home module

8.4.2 Showroom

The Showroom module, shown in Figure 8.7, is the focal point for extending the N-Gage experience. From here, players can read about new game releases in the Latest Games section.

Figure 8.7 The Showroom module with media and review sections

Before downloading a game, users can find plenty of information about it, including details such as category, number of players, publisher, release data, and age rating. They can also see game screenshots and video trailers, and read the ratings and reviews given by other players.

All N-Gage games also offer trial content, so that the users can see what the game is about before paying for it. The introductory levels are made available free to give a taste of what playing the game would be like. The user can then buy a license to get access to the entire game. The license is purchased from within the game in four steps. The N-Gage platform connects securely to the retailer backend to complete the transaction and retrieve a valid DRM license from Nokia's license server. The purchase steps are as follows:[4]

1. **License Selection**: The N-Gage DRM solution permits users to buy a limited or unlimited license for a game. Each license provides complete access to the game; however the limited licenses will expire after the timeframe specified at time of purchase. The unlimited 'full' license has no expiration. This step allows the user to select the license type they want to buy.

2. **Payment Option Selection**: N-Gage offers different payment options depending on the content, the license, as well as the buyer's region and operator. Payment by major credit card is always available. This step gives the user the opportunity to select the payment option and to enter their details, if necessary.

3. **Purchase Confirmation**: Before the transaction is processed, the user is asked to accept the terms and conditions, provide an email address for obtaining purchase receipt, and confirm the purchase details.

4. **Receive Activation Code and Retrieve License**: The transaction is confirmed and the user receives an N-Gage activation code, which provides proof of purchase. The user connects the phone to the license server by going online, and enters the activation code received. This enables the phone to retrieve a valid DRM license for the game.

The game is now activated and ready to play!

8.4.3 My Games

The My Games module is a collection of the games, trials and demos the user has installed, and provides quick access to start a game, rate and review it, or uninstall it.

---

[4] In some countries, users will be able to get game trials and purchase N-Gage activation codes directly from physical retailers. Once they've bought the code at a store, they can skip the first three steps and input the purchased N-Gage activation code directly, as described in step 4.

A user can get ratings and reviews from other players before making the decision to download a game. In the My Games module, the user can rate a game with 1 to 5 stars and submit review comments for it. The ratings and comments are uploaded online and made available to the other players in the N-Gage community. Users must create N-Gage accounts in order to submit their ratings and reviews.

N-Gage users can also suggest favorite and recently discovered games to their friends using game recommendations. Game recommendations are sent from the My Games module, and allow users to send a message about a particular game to one or many friends.

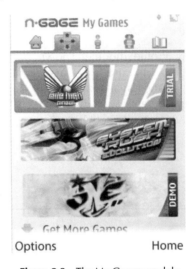

Figure 8.8   The My Games module

Figure 8.8 shows the My Games display. The games are shown differently according to their current status, as listed below. By default, games are sorted by last-played date (most recent first).

Owned Games: Owned games are those for which the user has purchased some type of license. The license may be time- or content-limited, but as long as the license is valid, full games appear with no restrictions and with full visibility of the N-Gage Points gradient bars.

Trial Games: Full games packaged with a limited license (time or content-wise), designated with a pink 'TRIAL' banner, shown on the right hand side of the display.

Demo Games: Demos are smaller sections of the game (available prior to game release) that show only a very limited feature and level set of the game. They are designated with a purple 'DEMO' banner, shown on the right hand side of the display.

Expired Games: Expired games are those for which the purchased licenses have expired.

Not Installed Games: Games that were once installed but later uninstalled, or games which have been downloaded or have been embedded in memory cards, but have not yet been installed. A game that is installed on a memory card is shown as 'not available' when the memory card has been removed, by being grayed-out.

Whenever a new game is downloaded or transferred over USB, the N-Gage application guides the user through installation, offering the different drives available, where they have sufficient memory, and giving the user the opportunity to choose which one is used for installation.

8.4.4　My Profile

The My Profile module, shown in Figure 8.9, tracks and stores key details about the user's N-Gage game history. Users can customize their profile, set their availability to play, and view game rankings and in-game accomplishments, including detailed N-Gage Points earned in gameplay and for community activities.

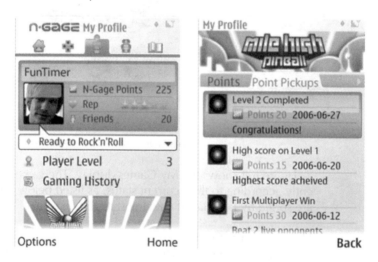

Figure 8.9　The My Profile module

All N-Gage games award N-Gage Points for completing certain in-game goals, called Point Pickups, as defined by the game. Games can award up to 1000 N-Gage Points total. The user's points earned, and details on each Point Pickup achieved, are recorded in the N-Gage Profile and can be used to compare with friends and other players in the N-Gage community.

There are three different types of N-Gage Points:

- **Solo Points**: obtained by playing through the game as a single player.

- **Multiplayer Points**: obtained when using the online features of the game and playing with others.

- **Arena Points**: obtained by participating in many of the community online activities, such as tournaments, events, and more.

Gradient bars on the right side of each game icon in the My Games module express the percentage of N-Gage points collected by the user, reflecting in an approximate way the user's progress in the game.

8.4.5 My Friends

The My Friends module, shown in Figure 8.10, is a player's link to friends in the N-Gage experience. From here, players can engage in one-on-one IM chat, send private messages, view each others' presence status and rate other players. For each friend in the Friends List, the user can see:

- presence (logged off/online, and available to play/online, and unavailable)

- number of N-Gage Points

- last played games.

Figure 8.10 The My Friends module

Users can easily invite other players from the N-Gage community to become their N-Gage friends by sending them a Friend Invitation, shown

Figure 8.11 Friend Invitation, rating, and chat in the My Friends module

in Figure 8.11, which can be customized to give a more personal touch to the invitation.

Users can set their own availability from any module in the application (available under the Options menu). Availability is defined by presence state, which the user can complement with a customized presence message to explain his or her current context.

Once online, the user can send private messages to other friends online, as well as to those who are offline. The messages will be stored on the server until the friend retrieves them. If online, users will receive notification of incoming messages while playing the games, to know when their friends are trying to reach them.

8.5   Becoming an N-Gage Developer

The previous section will almost certainly have excited you to the new possibilities for playing great mobile games on S60. The N-Gage platform makes it easy to find, purchase, share, and install games; a combination of distribution methods that makes it easier than ever for the consumer to acquire new games.

By this point in the chapter, if you are a game developer, you are probably also interested in the possibilities the N-Gage platform opens up. Nokia is making it possible to create high-quality native games using standard C/C++ idioms, and guarantees to eliminate device fragmentation and ensure game compatibility across a range of S60 smartphones. It also provides an end-to-end distribution mechanism; a dedicated gamer community for sharing game recommendations, chatting, and building a player profile; and multiplayer game support through the N-Gage Arena. So who is working on N-Gage games? What is it like to work with N-Gage? Can anyone develop games for the platform?

If you are a professional game developer, and work with a publisher that has a relationship with Nokia, you may already be working on an N-Gage game. At the time of writing, Nokia has revealed that the following publishers are creating games, and that new partnerships will be announced on an ongoing basis:

- Capcom
- Digital Chocolate
- EA
- Gameloft
- Glu Mobile
- Indiagames
- I-play
- THQ Wireless
- Vivendi Games Mobile.

Nokia will itself continue to publish game titles for the N-Gage platform. Some of the first generation N-Gage titles announced, such as **Mile High Pinball, Pro Series Golf** and **System Rush: Evolution**, have been created by developers working with the game publishing team that worked on the original N-Gage game deck titles.

Game developers working on N-Gage games sign a non-disclosure agreement with Nokia. Once their project has been authorized, they receive software (such as the N-Gage SDK) and hardware (including S60 smartphones) for reference testing. Registered N-Gage developers are also supplied with accounts to allow them to access the N-Gage Zone on Forum Nokia PRO. This site is primarily a developer resource maintained by the N-Gage developer support team. It provides up-to-date documentation specific to the creation of N-Gage games, from game design through to coding. There is also a set of documentation available to describe how to submit the completed game for certification; the detailed standard game requirements (SGRs) against which the game is certified are also published on the Zone, so that developers can familiarize themselves early on, and pre-test the game in advance of certification by Nokia.

The N-Gage Zone also gives developers access to the online support tool for requesting technical support, and a set of discussion boards where N-Gage developers can discuss technical issues with peer game developers in other companies, working on other titles.

N-Gage developer companies are assigned an account manager in Nokia, who assists them throughout the process of creating a title. They also receive support from a technical consultancy group within N-Gage developer support. Technical consultants visit the developer sites to provide training and answer questions about the N-Gage SDK. The technical consultant assists the team from the beginning of the project through to the final stages when the game is certified.

Currently, if you are not already working in a professional development team to create an N-Gage game for Nokia or one of the publishers listed above, you will not be able to get access to the N-Gage SDK. This was also the case for the original N-Gage game decks, as we described in section 8.1.3. However, as we mentioned in section 8.3.1 it is possible that many of the libraries found in the N-Gage SDK will become more widely available for developers outside the N-Gage program.

8.6 Summary

The evolution of the N-Gage game decks into the N-Gage platform of today shows what is possible for native mobile games on Symbian OS.

For developers, the N-Gage platform offers a way to write a game that runs across a range of S60 smartphones that offer the capabilities to run rich-content games. Device differences are abstracted, so a game is portable across the range of N-Gage compatible phones, and the idioms of C++ on Symbian OS are hidden behind a layer that provides professional-quality APIs for game development. Furthermore, the SDK features tools and APIs to hook the game to the end-to-end delivery system as well as the community provided by the N-Gage application and N-Gage Arena.

For users, through the N-Gage platform, Nokia aims to provide an easy, reliable service for users to discover, try, buy, play, share and manage mobile games. The next generation of mobile games is here. Get out and play!

Part Four

Java ME, DoJa and Flash Lite on Symbian OS

9

MIDP Games on Symbian OS

Sam Mason
(Mobile Intelligence)

9.1 Introduction

In this chapter, we explore the mainstay technology of mobile games – Java ME (the ME stands for micro edition). Java ME has led the way in mobile application and game development from its inception and has been the success story of mobile game development to date. We will start with a very brief Java ME history and discuss some of the features that differentiate the Symbian OS Java ME implementation from mainstream devices.

We'll then have a quick look at why you might use Java instead of Symbian C++ for game development, some architectural designs commonly used and then walk through a complete game example. We'll then wrap up with a look at the new SNAP framework from Nokia, that provides an infrastructure to support, standardize and streamline multiplayer games for MIDP.

Space is *really* tight, so we're only going to lightly touch on a number of areas of the technology. If you're not familiar with Java ME, the Java Community Process (JCP) and Java Specification Requests (JSRs), check out *en.wikipedia.org/wiki/JSR* and *java.sun.com/javame/index.jsp* for a good introduction. In addition, there is a ton of books written about Java ME development, which I've listed in the Further Reading section at the end of this chapter, so we aren't going to discuss high level GUI application development here. Instead we're going to focus on using the Java ME platform for game development.

Let's start with a quick history lesson...

9.2 Java ME 101

You're probably wondering just why we have a whole chapter devoted to Java ME in this book. The answer is threefold – mobile games are the killer applications of the wireless industry and have been for almost a decade, almost all of them were written in Java and there are over 1 billion Java-enabled handsets in the world. It's a great technology in all its forms, it's secure, mature, and even the name of the language is cool – 'Java' (just say it out loud). Over 5 million developers just can't be wrong.

Not only is it the best cross-platform mobile development technology in the universe, it's also really easy to write mobile games using it. Pretty much every mobile phone handset sold anywhere over the last four years (including *every* Symbian smartphone) has had built-in support for game development using the Java ME platform in the form of the MIDP 2.0 Game API. Check out a series of papers on runtime environments by Roy Ben Hayun, published in 2007 on the Symbian Developer Network (**developer.symbian.com**) if you need more information.

Devices sold prior to 2003 generally don't have this support (so if you have one of these, it's time to upgrade your hardware), although that didn't prevent a huge number of titles being developed as the early pioneers of constrained device development came up with a multitude of workarounds and *de facto* standards. This led directly to the widespread acceptance, market penetration and overall success of Java ME (or J2ME as it was known at that time) and the ongoing viability of using it as the game development platform of choice.

Sun Microsystems recognized back in the early 1990s that Java would be a major player in the resource-constrained market and subsequently launched an R&D project called Spotless to address this. The result was a small Java virtual machine (JVM) that used less than 10 % of the standard desktop version (160 KB to 512 KB) and was called the Kauai virtual machine (KVM – note that the 'K' does not, as is commonly held, stand for 'Kilobyte').

Since there's such a diversity of electronic devices with limited capabilities, Sun wasn't just thinking about mobile phones. Their vision included electronic toys, household appliances (like kettles), pagers, set-top boxes, digital TVs, personal organizers, and PDAs. Consequently, Java ME was divided into three main building blocks to address this diversity.

9.2.1 Configurations

A configuration represents a particular horizontal category of devices that share similarities in terms of memory budgets, processor power, and network connectivity, and can be thought of as defining their lowest common denominator. Configurations have a virtual machine and a set of classes for basic functionality only.

There are currently two configurations, the Connected Limited Device Configuration (CLDC), which was aimed specifically at mobile phones, two-way pagers and low-level PDAs, and the Connected Device Configuration (CDC) which was aimed at devices with more memory, better network connectivity, and faster processors, such as set-top boxes, high-end PDAs and communicator-type phones such as the Nokia 9300 and 9500.

As we are focusing on mobile phone game development, we won't say anything more about the CDC in this book. The CLDC, which is what most concerns us, is the configuration found on most mobile phones today. Most of the handsets sold in the last 3–4 years support the second incarnation of the CLDC, which is version 1.1. It can be regarded as an incremental update and simply added some new features to the configuration, including (but not limited to) thread naming and interrupt support, weak references and support for floating point operations (the `Float` and `Double` classes). As a result of the last addition, the minimum total memory requirement was also lifted from 160 KB to 192 KB.

9.2.2 Profiles

Profiles sit on top of a configuration and allow it to be adapted and specialized for a particular class of devices in a vertical market such as mobile phones. They also generally map to consumer expectations of portability – you may expect a mobile game to be portable across a set of mobile phones, but you wouldn't (necessarily) expect it to run on, say, your kettle.

A profile effectively defines a contract between an application and a set of similar devices. So if there *was* a Kettle Profile (KP), then coffee-making applications written using it would be expected to work on all KP compliant kettles – and *not* on mobile phones.

There are currently only two profiles that use a CLDC as a foundation – the mobile information device profile (MIDP) for mobile phones and the Information Module Profile (IMP) targeted at hardware without rich graphical displays, such as parking meters, home alarm systems and vending machines. CLDC is also used by the proprietary DoJa profile for i-mode devices from NTT DoCoMo (the subject of Chapter 10, which discusses writing games for the Japanese market). There are also a number of profiles built upon the CDC, but we won't cover them here, because in this chapter, we're only interested in MIDP.

The MIDP defines the core functionality that mobile applications can leverage – the UI, local storage, network connectivity, and application life cycle management. The original version 1.0 was started in November 1999, and, even though it did not actively provide game support, literally hundreds of games were written for it nevertheless. It was so successful that, almost immediately, work began on defining MIDP 2.0.

The MIDP 2.0 GUI library consists of a high level API with support for standard widgets (textboxes, check boxes, buttons, etc.) and a low level API supporting graphics contexts aimed mainly at game development. The low level API includes standard routines for creating graphics primitives (lines, polygons, arcs, etc.) as well as supporting clipping, full-screen canvases, regions, image manipulation, fonts, brushes, colors, and off-screen buffers.

Crucially, MIDP 2.0 also introduced timers and the Media API (a subset of the optional Mobile Multimedia API or JSR135) which are obviously relevant to our game focus, as well as the Game API. We'll talk more about that in just a bit.

9.2.3 Optional Packages

These are general purpose APIs in the sense that, while they provide a specific set of functionality, they do so for a potentially wide range of devices. Good examples of this include the Mobile 3D Graphics for J2ME (JSR184) and the Location API (JSR 179) both of which provide functionality that is not specific to a family of devices such as mobile phones. Optional packages effectively add functionality to a profile and may even end up incorporated into one as time goes on and the technology matures.

Java's cross-platform ideal has always been the ability to 'write once, run anywhere' and the language was specifically designed with that end in mind. Unfortunately, optional packages have often turned out to be a double-edged sword. While allowing relatively fast dispersion of new technologies into the marketplace, they have resulted in what is generally known as 'fragmentation.' This is a nice word for it – and for a long time what it really meant was that you could write a Java ME application only once as long as you only wanted it to run on one type of handset.

In reality then, people often had to re-write or re-design many times to accommodate missing APIs, implementation variations, and workarounds to address known issues. Sometimes API implementation details vary even within the same family of devices (or in the worst case, each device). Sadly the phrase 'write once, debug everywhere' was really a lot more apt. In general, this had the effect of greatly hampering the time to market of Java ME applications and games.

9.2.4 Fragmentation

There have been two efforts to address fragmentation in the past. The first was the Java Technology for the Wireless Industry (JSR 185). The JTWI aims to minimize fragmentation across handsets by defining a minimum level of support for optional APIs. Release 1 of the JTWI, approved in July 2003, defines three categories of JSRs – mandatory, conditionally required, and minimum configuration. A handset that declares itself to be JTWI-compliant supports at least the following mandatory JSRs:

- MIDP 2.0 (JSR 118)
- Wireless Messaging API (JSR 120).

The minimum configuration is unsurprisingly CLDC (either version). If the device allows MIDlets to access multimedia functionality via APIs, then the JTWI also specifies that the Mobile Media API (JSR 135) is required (i.e., it is conditionally required).

It also defines a number of implementation parameters for the JSRs, including things like support for JPEG encoding, MIDI playback, the number of record stores allowed in the RMS, Unicode support, and some threading requirements.

In December 2006, this was taken a step further with the release of the final draft of the MSA – the Mobile Services Architecture (JSR 248),[1] which can be thought of as JTWI 2. This mandates that an MSA compliant device is JTWI compliant as well as providing support for the following APIs:

- Security and Trust Services API (177)

- Location API (179)

- SIP API for Java ME (180)

- Content Handler API (211)

- Payment API (229)

- Mobile Internationalization API (238)

- Multimedia Supplements (JSR 234)

- Web Services (JSR 172)

- Scalable 2D Vector Graphics API for Java ME (226)

- **Messaging (JSR 205)**

- **Mobile 3D Graphics API (JSR 184 1.1)**

- **Mobile Media API (JSR 135)**

- **Bluetooth (JSR 82)**

- **File and PIM (JSR 75)**

- **MIDP 2.0 (JSR 118)**

- **CLDC1.1 (JSR 139).**

The specification defines two categories of MSA known as MSA and MSA Subset. A device that is MSA Subset compliant supports only the last seven (bolded) JSRs from the list above. This is a huge step forward, but at the time of writing there are less than a dozen or so

---

[1] You can find out more about JSR 248 at: *jcp.org/aboutJava/communityprocess/final /jsr248/index.html.*

handsets in the market that are MSA compliant, and they're not actually Symbian smartphones (you can find the latest list of compliant devices at ***developers.sun.com/mobility/device/device***). So it will take a while until we can rely on this level of standard support.

Ok, that's about all we need to discuss in the history of Java ME, particularly as there are many excellent textbooks that cover this in depth. See the references at the end of the chapter for more information.

9.3   To Java ME or Not To Java ME?

Some people will tell you that Java ME (and actually Java itself) is too slow for any type of software development, let alone game development. Well, they're wrong. First and foremost, if you want your game to run on as many mobile phones as possible, you need to use the most ubiquitous platform in existence. Second, there is a huge support base for Java ME both in industry and from developer networks. Third, a well designed MIDP game can be ported across to DoJa fairly quickly which opens up the possibility of entering the largest game market on the planet – Japan. I'll discuss some of the issues involved in a port to DoJa in Chapter 10.

The ability to perform rapid application development in the mobile space is imperative given the short time to market demands of today's games industry. As much as I love C++, I'll still turn to Java ME almost every time for quick prototypes. Especially when you can demo it on Symbian smartphones, BlackBerry phones, Nokia Series 40 handsets, or even one day maybe an iPhone (if they let us), and on independent proprietary OS phones from Fujitsu, LG, Telstra and so on.

There's a large number of other reasons to use the Java ME platform for game development as well. To start with, there are millions of skilled Java developers worldwide and the learning curve from standard edition Java to Java ME is actually quite flat. This means obtaining quality professionals is easy for game development studios. In addition, there are forums and hundreds of websites with code samples, tutorials and techniques. These weren't written by academics or corporations, but by pioneers who had to work out how to actually make this technology work in the early days.

The Game API, introduced in MIDP 2.0 addresses many issues realized in the MIDP 1.0 days by introducing techniques such as double buffering, a key state input mechanism that allows for multiple keystroke detection and a high-level mapping, allowing the abstraction of main game actions from specific keys. Compared to game development using Symbian C++, this is a lot easier to do, and there are more people available to do it. The Game API also introduced sprites, tiles and layer management support. Sprite transformations, (which allow for reduced JAR sizes), image transparency and collision detection at a Sprite and pixel level are also included, which allows the developer to focus on the game logic for a better product.

So these are some of the reasons why you should use Java ME for game development. Now here are some reasons why you should be careful before you do:

- even though advanced hybrid techniques are used to produce the final executables, Java ME applications are just not going to be as fast or efficient as a well written native Symbian C++ application. That said, Symbian engineers have put a large amount of time into getting Java ME applications to run as fast as possible with a huge amount of success, as evinced by internal benchmark tests.

- you are limited to the functionality exposed via the CLDC, MIDP and Optional APIs. While there are some ways (hacks) to access native functionality, they're architecturally poor, not a lot of fun and are usually best avoided – in short, if you need native, *be* native and work in C++

- fragmentation across handsets, which has already been discussed

- updates to the JavaME profiles and JSR support can be very slow, so if you're dependent on a particular API, it may hold up your product

- many handsets have quite restrictive limitations on core application parameters such as a maximum allowed JAR size, maximum heap memory, as well as limitations on available memory for data persistence. For larger database style applications, these can be prohibitive

- MIDlets can be slow to load due to the need to load the entire VM binary and AMS (if not already resident)

- security built-in via bytecode verification and a sandbox security model. This is independent of (but also built on) the new platform security model introduced in Symbian OS v9.

However, as we'll see in the next section, many of these issues don't apply on Symbian smartphones as their offering is best of breed.

9.4 Tools and SDKs

Getting started in Java ME development is really easy and there are a number of great tools and IDEs freely available online to help get you going:

- Eclipse with the EclipseME plugin (***www.eclipseme.org***)

- Netbeans using the Netbeans Mobility Pack (***www.netbeans.org /products/mobility***)

- Sun Wireless Toolkit (WTK) (***java.sun.com/products/sjwtoolkit /download.html***).

Until recently, Nokia also supplied Carbide.j,[2] which I liked a lot, and it plugged itself into Netbeans as well. However, support for this has now been withdrawn, as apparently there's not a lot of point having it around when so many people can get Eclipse or Netbeans for free. Nokia does have a large number of SDKs for MIDP development for S60 and Series 40 devices though and both Eclipse and Netbeans will let you use these (and switch between them) quite easily in your projects.

Although I've used Netbeans happily for years, these days I tend to favour Eclipse in the interests of IDE consolidation – apart from using it to build Java ME software for Symbian OS, you can also use it for BlackBerry development (with some work) and, as we'll see in the next chapter, you can use it to target i-mode devices in the Japanese market with the DoJa plugin from NTT DoCoMo. In addition, Carbide.c++ is based on Eclipse, so the look and feel is the same, whatever work you're doing.

If you've never done any Java ME work before, the WTK from Sun is the best place to start. It's a simple development tool that allows you to focus on learning Java ME technologies without having to learn the intricacies of a new IDE up front. The latest version of the WTK is 2.5.1 and it also includes the SNAP Mobile SDK which is used for providing standardized community features in online game development. This is something that we'll cover in section 9.10.

9.5   What About MIDP on Symbian OS?

Let's jump straight into why Symbian's implementation of CLDC and MIDP provides one of the best game development platforms in the mobile space. Symbian first added CLDC 1.0 and MIDP 1.0 support in Symbian OS v7.0 in 2002. The CLDC was a port of Sun's CLDC 1.0 Hotspot Implementation VM (CLDC HI). This was about ten times faster than the standard KVM available at the time by virtue of leveraging techniques such as lightweight threading and dynamic adaptive compilation. Then from Symbian OS v8.x onwards, they supplied a port of CLDC 1.1 HI, thus adding floating point support for Java ME applications to their offering.

Then came the unforgettable Nokia 6600. Based on Symbian OS v7.0s, it was the first Symbian smartphone to support the MIDP 2.0 profile and most of the JTWI release 1. The main point to note here is that in this case, MIDP 2.0 was actually implemented by Symbian engineers and this allowed them to leverage the underlying operating system services and interfaces to maximize performance. Both Symbian and Nokia (who

---

[2] ***www.forum.nokia.com/main/resources/tools_and_sdks/carbide/index.html***

implemented most of the optional JSRs on S60 handsets) were able to realize them as interfaces to native OS services which minimized the footprint, while providing MIDP tight integration with the OS itself. This allows MIDP applications on Symbian OS to use the native widgets for the UI and thus they can be hard to differentiate from C++ applications. This is great, as users often switch back and forth between multiple applications, and they should maintain the same look and feel.

Symbian OS is a full multi-tasking operating system, and the Java VM itself runs in a separate process in its own thread (the interpreter has its own C++ thread where native interface code also executes). In addition, there are several other native threads in the VM process and several more in the AMS. However, Java threads themselves do not use native threading – they are instead implemented by the virtual machine itself as lightweight threads in the CLDC HI.

Another consequence is that, as a multi-tasking operating system, Symbian OS doesn't need to automatically pause a 'backgrounded' MIDlet (one that has been sent to the background). Depending on your application this may or may not be a good thing. As a game developer, you'll certainly want to address this as best practice dictates, pausing your game loop and releasing as many resources as you can in this situation. Furthermore, this behavior is dictated by the Application Management System (AMS) which can be (and is) customized by licensees resulting in variance of behaviors between S60 and UIQ. We'll discuss more on how to handle this and the MIDlet life cycle later.

Other benefits Symbian OS offers in its MIDP implementation include a `Canvas` class that is double buffered by default, support for the native color depth of the underlying platform, and a wide range of network and push registry protocols. Within the bounds of available resources, Symbian OS does not restrict the size of JAR files, the number of RMS records or their size, and you can create as many socket connections as you want. Further, you aren't limited in how many Java threads you can create and the VM supports a dynamic heap so it can grow to meet demand. You need to keep that in mind when load testing your games as `Runtime.freeMemory()` will only return the instantaneous memory with respect to the current size of the heap.

All releases of Symbian OS starting with version 8.0 support the full Mobile Media API (JSR 135) and natively support OpenGL ES 1.0. OpenGL ES forms an abstraction layer for the Mobile 3D Graphics API (JSR 184), meaning that your MIDlets automatically benefit from any hardware acceleration on the handset without your having to do anything. This is way cool. And even though only a handful of Symbian smartphones currently support hardware accelerated graphics, this is just the start of a huge revolution in mobile applications and games.

Symbian OS v9 has enhanced a number of existing APIs and added many more. The net result is a rich diversity of functionality optimized

Table 9.1 Symbian OS API Support for the Java
ME Platform

API	JSR
Web Services	172
Security and Trust Services	177
SIP	180
Scalable 2D Vector Graphics	226
Location	179
File Connection & PIM	75
Mobile 3D Graphics	184
Mobile Media	135
Wireless Messaging	120/205
Bluetooth	82
Java Technology For The Wireless Industry (JTWI)	185

for performance and memory efficiency. The latest Symbian smartphones
support the APIs listed in Table 9.1.

If you're confused about the Wireless Messaging APIs (JSRs 120 and
205), don't be – you only need to know that JSR 205 is a superset of JSR
120 and specifically adds programmatic support for MMS.

One last point – one API that is notably absent from Table 9.1 is
the Java Binding for the OpenGL ES API (JSR 239). However it was only
finalized in September 2006, and, at the time of writing (September 2007),
it is currently undergoing a maintenance review. It is to be hoped that we
will see support for this in Symbian OS in the near future as well, since this
allows Java ME developers to code in the same immediate mode style used
in standard OpenGL ES. Consequently, porting native OpenGL/OpenGL
ES applications across to Java will be much easier as well.

9.6   Pausing for Breath

Even phones get tired. When power levels drop or memory becomes
scarce, the operating system will sometimes direct the AMS to pause or

even stop and destroy a running MIDlet altogether. Whether you like it or not, any game you write is going to get backgrounded on a regular basis. This happens frequently in response to operating system events, other applications moving into the foreground, or incoming messages/phone calls during game play. Sometimes it just happens accidentally because the player is unfamiliar with the phone's abilities and interface. However it happens, it's going to happen.

So, before you even consider how to start a game, think about how to pause it and how to stop it. This is really easy to get wrong, but with some planning and design it's actually quite easy to get it right, which directly impacts battery life and therefore game life. No-one wants to play a game that flat-lines their battery every time it's accidentally left running in the background.

By the nature of their ergonomics and the general use cases, mobile games are often played in short bursts – sometimes less than a couple of minutes – if that long. As Chapter 1 described, research shows that a large proportion of users play mobile games for very brief spurts of entertainment – usually while waiting for a bus, while waiting to buy a coffee, while standing in a queue, or even while waiting for their partner to get ready to go out! Furthermore, it should be remembered that often a game is paused by the player in response to a change in the player's surroundings – because they've just got on the bus, or entered a meeting, a lecture, or a job interview. So pausing a game needs to do just that – pause the game world entirely.

This usage pattern is completely different to that of PC games, and therefore we have to design the game architecture according to a different perspective. A mobile game will be paused and resumed frequently and given the constrained nature of mobile devices, it's important to make sure that the following steps are taken (in no particular order):

- stop all non-essential threads – this generally refers to the game loop but also includes timers and any other threads that the game uses.

- release any non-essential resources – references to any large objects, cached game data, handles to the onboard camera(s) or the microphone, and so on

- cease playing any video, sounds, and music

- update the display to reflect that it is in a paused state

- save the current game state.

In fact so important is this approach, that Chapter 2 covers this in some detail. The first two points above make sense for any mobile application not just games, but the third point, to cease playing any audio or video, is an interesting one. I have played more than one game that continues to play tones or audio tracks even when in a paused state – which can be very strange and results in a poor user experience.

This is often because such games are designed for more mass market devices that do not support multi-tasking and it is assumed that a MIDlet is either running or dead – so there is no concept of a 'paused' state. For example, on Nokia Series 40 devices, only one MIDlet can be active at any one time, although a MIDlet can be backgrounded by a system screen (such as an incoming call dialog). However the AMS will not call any MIDlet life cycle methods when this occurs – although recent events indicate that Series 40 may soon allow the execution of multiple MIDlets simultaneously.

As we have seen, this is not the case on Symbian OS. A MIDlet that has moved to the background will continue to run, consuming clock cycles of the processor, using resources, and draining the battery quite happily unless told otherwise.

It's also important to think through the last two points in the list above – what does a MIDlet display when paused, and how will a player resume game play? How you choose to do this depends on the actual game, but common techniques include displaying:

- a separate pause menu with a pre-selected 'Resume' option

- a 'Paused' caption over the top of a screen grab of the game world taken at the point where the game was paused

- the main menu with a 'Resume' option added – which should be at the top and pre-selected.

If you do implement a separate pause menu, it's also good practice to include menu items for accessing the game settings and for exiting the game altogether. Players should be able to pause a game, disable vibrations, music, and sound effects via the settings screen, and then resume play quickly and seamlessly in context with their surroundings. This is a real issue – a number of my friends have complained about being caught out in meetings because the game pause/resume cycle re-enabled sound effects and music without their realizing it (although they probably should have been working anyway). Many games do not handle pausing well, which is why Forum Nokia has published a white paper devoted entirely to discussing how to do it.[3]

---

[3] The paper is called *At The Core Of Mobile Game Usability: The Pause Menu* and can be found in the Usability section of the Forum Nokia website (**www.forum.nokia.com/main /resources/documentation/usability/**). The exact location of the paper has a URL which is far too long to type in from a book, and is subject to change anyway, if the content is updated in future. We will keep a set of links to papers and resources that are useful to game developers on the Symbian Developer Network wiki (**developer.symbian.com/wiki**).

MIDlet pause behavior is dictated by the application management system (AMS) which is itself customizable. So it turns out that under S60, the AMS does not call `pauseApp()` on a MIDlet when it moves into the background, whereas on UIQ it does (or should).

The best solution here is to ensure that all of your game loop pause code is centralized in a method that can be called as required in response to a variety of events. This includes the AMS pausing the MIDlet via `pauseApp()` (if it does so), the MIDlet notifying the AMS that it would like to be paused via `notifyPaused()` and finally by leveraging the `Canvas.showNotify()`, `Canvas.hideNotify()` and the `Displayable.isShown()` methods.

When resuming game playing, you can use the saved game state to quickly re-initialize all game data as required. You can even leverage this so that when the user next starts the MIDlet, it offers them the option to resume their last game which is a nice feature to have.

It's also a good idea to give the player a few seconds to re-orientate themselves in the game world before resuming the game loop. Not only does this give them time to position their fingers on the keypad and mentally prepare for the event, but this time can be used behind the scenes to re-acquire resources and re-initialize the game state.

Most games need to track time – whether this is for the game AI, the simulation based on your physics model (say fading out sprites in a particle system), or for more mundane tasks like ensuring a steady frame rate. Sometimes it makes sense to run your AI at a different rate to your simulation, depending on what you're doing. In any case, elapsed time is how we make our game worlds dynamic, independent of user input.

This is where the pause/resume process can trip you up if you're not careful. Mostly developers use `System.currentTimeMillis()` to monitor time in a game loop and ensure a steady tick rate. This is fine, but your game needs to work in virtual time (i.e., time should proceed smoothly as far as the game is concerned). It can get a bit daunting unless you're aware of this beforehand but, again, planning for the game pause makes this trivial. Simply reset your clocks by the amount of real time that has elapsed during the pause, and as far as the game world is concerned, it never happened.

9.7 Living it Up

When you start designing a mobile game, the very first thing you should be thinking about is how to leverage the MIDlet life cycle correctly. Most text book examples get this wrong and that's not all that surprising as there are some common misconceptions in the industry, mainly to do with delegation of responsibilities between the AMS and a MIDlet.

The AMS governs the life cycle of a MIDlet and is therefore responsible for affecting transitions on a MIDlet's state as well as tracking it. It moves a MIDlet between the active, paused and destroyed states by invoking the `startApp()`, `pauseApp()` and `destroyApp()` methods of the MIDlet base class respectively. These methods are callback methods for the AMS and it is the responsibility of the developer to provide non-trivial implementations of them.

MIDlets can initiate state changes to the paused or destroyed states themselves and inform the AMS of this by calling either the `notify-Paused()` or `notifyDestroyed()` methods. A paused MIDlet can also let the AMS know that it would like to re-enter the active state by calling the `resumeRequest()` method.

The thing to remember is that your code should *never* call the AMS call back methods. It's very common to see code that does this like that shown below:

```
private void shutDown(){
  try{
    destroyApp(true);
    notifyDestroyed();
  }
  catch(MIDletStateChangeException e){
    // whatever
  }
}
```

While this will work, it assumes that the cleanup code in a MIDlet initiated shutdown is the same as that in an AMS initiated one. Further, this can lead to inconsistencies – what if the OS directs the AMS to destroy the MIDlet while the above method is executing? Then the `destroyApp()` method gets called twice as does any cleanup code it contains – which is a sub-optimal state of affairs to say the least. An analogous situation exists with AMS versus MIDlet initiated pausing.

The general solution here is to recognize that the MIDlet needs to track its own state internally to prevent duplicate code executions and separate out start, stop, pause, and resume code from the six life cycle methods. The other benefit of doing this is that it then becomes trivial to correctly handle automatic pause/resume functionality for the game when the MIDlet is moved in or out of the background for any reason. However, a word of caution is in order here – the state transition code needs to be synchronized, since it can be called by two different threads (the AMS and the MIDlet). However it's generally not good practice to make the AMS wait on an object lock as it may need to act immediately in response to a directive issued by the operating system (such as a low memory condition for example). Symbian has put a lot of thought into this problem though, so it won't cause any issues here.

Furthermore, this is done using the same centralized methods used in handling the life cycle state transitions, as the following code snippets illustrate (adapted from *Programming the MIDP Lifecycle on Symbian OS*, a 2005 paper by Martin de Jode which can be found on ***developer.symbian.com***).

Managing the life cycle state transitions:

```
public class GameMIDlet extends MIDlet {
   ...
 private final static int PAUSED = 0;
 private final static int ACTIVE = 1;
 private final static int DESTROYED = 2;

 private int iState = PAUSED;
 ...
 public void startApp(){
   start();
 }

 public void startMIDlet(){
   resumeRequest(); // request startApp()
 }

 private synchronized void start(){
   if(iState == PAUSED){
     controller.startApplication();
     iState = ACTIVE;
   }
 }

 public void pauseMIDlet(){
   pause();
   notifyPaused();
 }

 public void pauseApp() {
   pause();
 }

 private synchronized void pause(){
   if(iState == ACTIVE){
     controller.pauseApplication();
     iState = PAUSED;
   }
 }

 public void exitMIDlet(){
   cleanup();
   notifyDestroyed();
 }

 public void destroyApp(boolean unconditional) {
   cleanup();
 }
```

```
private synchronized void cleanup(){
  if(iState != DESTROYED){
    controller.stopApplication();
    iState = DESTROYED;
  }
}
}
```

Handling background and foreground events:

```
public class GameScreen extends GameCanvas
             implements Runnable{
...
public void showNotify(){
  if(controller.gamePaused()){
    reactivateMIDlet();
  }
}

public void hideNotify(){
  if(controller.gameRunning()){
    pauseMIDlet();
  }
}
...
}
```

These concepts are summarized in Table 9.2. For a more detailed analysis, please refer to the de Jode paper, *Programming the MIDP Lifecycle on Symbian OS*, mentioned earlier.

9.8   Game Architecture

Software architecture is a contentious topic and generally a highly subjective one – after all if you put three computer scientists in a room you'll generally get six opinions on how to get out (and one of them will have at least three ways in mind). When you add in the large disparity of skill sets and handsets available in the mobile game industry, you find that well, not everyone agrees.

However, given that Java ME broke new ground in MIDP 1.0 days, it's fairly common to adopt design patterns commonly used in mainstream Java development (standard and desktop) such as the Model-View (MV) and Model-View-Controller (MVC) patterns.

Since most Java ME game development occurred before MIDP 2.0 was released, companies created their own custom game packages under MIDP 1.0 to promote re-use of their code base. In general, MIDP 1.0 devices had quite severe constraints on core application parameters such

Table 9.2 The six life cycle methods

METHOD	Called By	Conditions	Notes
`startApp`	AMS	MIDlet first started	
`startApp`	AMS	A previously backgrounded MIDlet in the PAUSED state is moved into the foreground	Can occur multiple times so need to flag if any one-off initialization occurs
`startApp`	AMS	When the AMS finds that it can service a `resumeRequest` call from the MIDlet	
`pauseApp`	AMS	Low battery conditions on a backgrounded MIDlet that is not in the PAUSED state	Automatically invoked when backgrounded, depending on UI platform (S60, UIQ)
`destroyApp`	AMS	User terminates the MIDlet manually	
`destroyApp`		Low memory conditions or power down	
`notifyPaused`	MIDlet	MIDlet voluntarily pausing itself	Tells AMS to update it's copy of the MIDlet state
`notify Destroyed`	MIDlet	MIDlet voluntarily destroying itself	Tells AMS to update it's copy of the MIDlet state
`resumeRequest`	MIDlet	Informs the AMS that the MIDlet would like to re-enter the ACTIVE state.	The AMS may not be able to immediately service this request due to current conditions

as JAR size, available heap memory, and small amounts of RAM. Part of this process included having to determine what architectures worked and what didn't under these conditions, and in some respects, these modest beginnings still guide (and to some extent even restrict) development processes even today.

However, it's now 2008 and given the short average lifetime of a phone handset, the point to remember here is that these restrictions generally don't exist in the new generation of smartphones – and in particular Symbian smartphones. So, many of the more cautious approaches, while valid, need to be re-assessed for relevance in the current market.

In many cases, we can now accept a larger JAR size for the sake of building a more structured class hierarchy than was previously viable. Rather than avoiding interfaces, directly exposing member variables, and optimizing string references to minimize class sizes, we can shift the design focus away from (necessary) minimalism across to consistency, re-usability and maintenance.

9.9   Case Study: *Third Degree*

In this section I'm going to run through the main points of the sample game, ***Third Degree***, that is supplied with this chapter. The game's splash screen and a screenshot from the gameplay are shown in Figures 9.1 and 9.2 respectively. You can download the code from this book's main page on the Symbian Developer Network at ***developer.symbian.com***. The game is a traditional 2D arcade-style game which embodies most of the techniques I've discussed here. I strongly suggest downloading the

Figure 9.1   Splash screen display

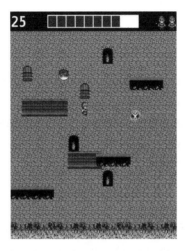

Figure 9.2 Action shot of our hero in mid jump

source code and getting familiar with it, as there's not going to be enough room here to cover it in depth. The point that I want to make here is to demonstrate *one* way of undertaking the game development process – and there are certainly others.

It's at this point where the author usually says "it's just a sample game for illustrative purposes." Well it's not; it's a complete game. While it certainly doesn't have all the features I wanted to put into it, I've tried to make it as functional and interesting as possible (although I'm a terrible graphic artist and an even *worse* sound engineer).

In this game, you're the hero who has to jump across a series of moving platforms over an endless pit of fiery death, racing the clock to reach safety (and presumably long life and happiness). You get points for successful jumps and bonuses along the way as well as negotiating the deadly Rings of Fire (which burn you and reduce your jump power). You get three lives to make it to the end and a time bonus if you do.

And if you haven't worked it out yet, if you miss or fall off a platform at any stage along the way, you die.

The process of building this game started off with a set of 10 fundamental questions. And there's no point even thinking about starting development if you can't answer these clearly at the outset:

1. What are the aims of the player?

2. How will pausing be handled (whether system or user initiated)?

3. What is the 'Game Over' condition?

4. What controls will be used for player input?

5. How will the game world be represented and simulated (is a physics or AI engine appropriate)?

6. How will the frame rate and simulation rate affect each other (if at all)?

7. How will changes in display size affect game play?

8. What media effects will be used (audio, video, vibra)?

9. How will scores, lives, bonuses and damage be handled and displayed to the player, e.g., through a heads-up display (HUD)?

10. What optimizations should be employed in the simulation and rendering phases?

I also made the following design decisions right at the start:

- to favor a stronger class design at the cost of a larger JAR file

- to limit the frame rate to keep consistent performance across devices

- to use a virtual simulation rate whereby a fixed number of game ticks represents one unit of time in the physics model

- that user input would be limited to move left, move right and jump actions

- to dynamically adjust the main menu to handle pauses (no pause menu)

- to use the MIDlet class only as an interface to the AMS and screen changes

- to combine the roles of simulation/physics engine and sprite (layer) management into a single class derived from the `LayerManager` class

- to use the MVC design pattern for the application framework, but keep the game sub-system separate.

By keeping the actual game sub-system separate, it's easier to cleanly delineate MIDlet life cycle events from game life cycle events. To harp on pausing (again), remember that pausing the game doesn't (necessarily) pause the MIDlet, but that pausing the MIDlet *should* pause the game (if it's running). Likewise, should stopping the game equate to stopping (exiting) the MIDlet? If you've kept the event handlers separate, that's a design decision that's easily changed later.

The high-level class diagram for **_Third Degree_** is shown in Figure 9.3. Note that only the main classes have been included for clarity.

The core of the game comes down to three main classes and the world objects themselves which I'll now discuss in turn, focusing on their high-level roles.

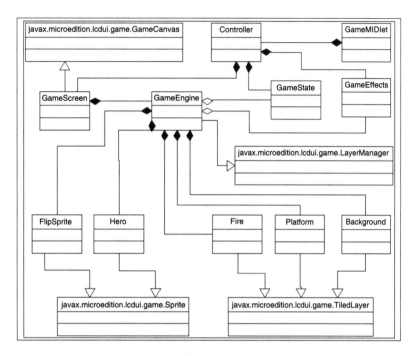

Figure 9.3 UML class diagram for ***Third Degree***

9.9.1 `Controller` Class

The `Controller` class links the outside world (the AMS and the MIDlet class) to the game sub-system. It serves as the conduit through which all high-level applications events are handled. It is responsible for monitoring and changing the state of the game sub-system in response to system, game, or user events.

It also holds an instance of the `GameState` class that contains the current score, health, and number of lives of the player. This can be persisted if necessary to allow a paused game to be re-started after the MIDlet exits or the device is shut off. The controller also creates the `GameScreen` instance upon which all rendering is performed.

```
public class Controller{
   ...
   private final static int GAME_STATUS_NONE          = 0;
   private final static int GAME_STATUS_INITIALISED   = 1;
   private final static int GAME_STATUS_PAUSED        = 2;
   private final static int GAME_STATUS_RUNNING       = 3;
   private final static int GAME_STATUS_STOPPED       = 4;
   private final static int GAME_STATUS_OVER          = 5;
   ...
   private GameMidlet iMidlet;
   private int iGameStatus;
   private GameState iGameState;
```

```
private GameScreen iGameScreen;
...
public void startApplication(){
  if(iGameStatus == GAME_STATUS_NONE){
    initialise();
  }
  if(iGameStatus == GAME_STATUS_PAUSED){
    resumeGame();
  }
}

public void pauseApplication(){
  if(iGameStatus == GAME_STATUS_RUNNING){
    pauseGame();
  }
}

public void stopApplication(){
  if(iGameStatus == GAME_STATUS_PAUSED
    || iGameStatus == GAME_STATUS_RUNNING){
    stopGame();
  }
}

private void resumeGame(){
  iGameScreen.resumeGame(iGameState);
  iGameStatus = GAME_STATUS_RUNNING;
  ...
}

private void pauseGame(){
  iGameScreen.pauseGame();
  iGameStatus = GAME_STATUS_PAUSED;
}
}
```

9.9.2 GameScreen Class

GameScreen creates the game engine and is primarily responsible for
executing the game loop. On instantiation and on a sizeChanged event,
it calculates an optimal layout for the world items in terms of the available
screen real estate. In terms of rendering, it draws the heads up display
(HUD) for the player data and delegates any remaining rendering to the
game engine (which, if you'll remember, sub-classes LayerManager). It
also captures user input via standard key stroke game mappings to control
the hero and handles commands to pause or resume the game loop.

```
public class GameScreen extends GameCanvas implements Runnable{
  ...
  public void run(){
    while(!stopped){
      processUserInput();
      integrateWorld();
      render();
```

```
      flushGraphics();
      synchroniseFrameRate();
    }
  }
  ...
  private void integrateWorld(){
    long currentTime = System.currentTimeMillis();
    long elapsedTime = currentTime - lastIntegrationTime;
    if(elapsedTime > GameEngine.MILLISECONDS_PER_TIMESTEP){
      gameEngine.integrateWorld();
      lastIntegrationTime = System.currentTimeMillis();
    }
  }

  private void synchroniseFrameRate(){
    long currentTime = System.currentTimeMillis();
    long elapsedTime = currentTime - lastFrameTime;
    lastFrameTime = currentTime;
    if(elapsedTime < GameEngine.MILLISECONDS_PER_FRAME){
      try{
        gameThread.sleep(GameEngine.MILLISECONDS_PER_FRAME - elapsedTime);
      }
      catch(Exception e){}
    }
    else{
      gameThread.yield();
    }
  }
}
```

As can be seen in the listing above, the frame rate is kept to a maximum value, so that the animations do not proceed unrealistically on faster hardware. This is done by sleeping the game thread if it has taken less than MILLISECONDS_PER_FRAME to complete the game loop. By setting this to 40 ms, we ensure a rate of about 25 frames per second, which is plenty.

We also track how often we integrate the world to effect the simulation. This should *not* occur at the same rate as rendering (which is a common mistake). By using the MILLISECONDS_PER_TIMESTEP constant, we can tick the world simulation at a rate of our choosing to keep it smooth and realistic, minimizing what are often complex calculations.

9.9.3 GameEngine Class

Not surprisingly GameEngine's responsibilities include generating the world, scrolling through it in response to movement, and detecting collisions between world objects. It also handles all updates to the player's health, lives, and score.

Fundamentally, the game engine must run (physics) sequences to give the game a realistic effect – so that when our hero runs, falls, lands on a moving platform, jumps, or catches alight, the changes in movement and world state all look fairly realistic. The engine ticks all world objects whenever its `integrateWorld()` method is called from the `GameScreen`. It then checks for collisions or other events that must be handled (such as picking up bonus points).

```
public class GameEngine extends LayerManager{
  ...
  public static final int GRAVITY = 10;
  public static final int TERMINAL_VELOCITY = 50;
  public static final int FIRE_TICK_RATE = 2;
  public static final int BASE_RUNNING_SPEED = 10;
  public static final int JUMP_FORCE_NORMAL = 3;
  public static final int BURN_TICKS = 5;
  ...
  public void integrateWorld(){
    try{
      tickHero();
      tickWorldItems();
      updateWorldState();
    }
    catch(Exception e){
      e.printStackTrace();
    }
  }

  private void tickHero(){
    if(isJumping()){
      hero.setVerticalSpeed(stepJumpSequence());
    }
    else if(isFalling()){
      hero.setVerticalSpeed(stepFallSequence());
    }
    else if(isOnPlatform()){
      hero.setVerticalSpeed(platforms[currentPlatformIndex].getSpeed());
    }
    hero.tick();
  }
```

9.9.4 World Objects

The objects in this game consist of the `Hero`, `FlipSprite`, `Fire`, `Platform` and `Background` classes. Of these, the first two are specializations of the `Sprite` class, while the rest are derived from the `TiledLayer` class. `FlipSprites` simply oscillate between two states constantly to form a simple and cheap animation. They are used for bonuses, the Rings of Fire and health power-ups. The `Fire` class is a tiled layer that can be dynamically built to fill the available screen width, and the `Platform` is a tiled layer to allow for random sizes. These objects are built using the world object frames shown in Figure 9.4.

The `Background` class contains a number of artifacts and animated tiles. It is created using the image shown in Figure 9.5.

The `Hero` is implemented as a finite state machine with the engine responsible for effecting state transitions. This approach allows the hero's

Figure 9.4 World object frames

Figure 9.5 Background tiles

animation sequences to be switched in a well defined context-sensitive manner throughout the entire game. Note that, by design, world objects such as the `Hero` don't know anything about game physics and don't make any decisions for themselves. They're just sprites, so they really only need to know where they are, where to move to and how they should look on the way there.

9.9.5 Game Effects

The game uses a series of hard coded byte arrays as tone sequences through the `Game Effects` class. These are triggered by game events such as incurring damage, gaining health or bonuses, achieving death or finishing the game. Vibrations are also used when the hero falls onto a platform from a given height (so it won't always happen which is good as you don't want to drain the battery). Both of these effects can be turned off on the Options screen.

9.10 Did You Hear That SNAP?

Mobile game playing is a force to be reckoned with. As a concept, we're all aware that it has driven innovation, technology, and business over the last decade or so, but its effect, in terms of our societal interactions and expectations, is much more profound. We think nothing of spending hours online playing games with people that we'll never meet, in countries that we'll never visit. People are paying others to play online games such as ***World of Warcraft***, accruing virtual assets on their behalf, and anonymous entities known only by their handles can live in a world where their game skills reign supreme.

With trends like this, it's important to consider multiplayer game development in the mobile space. This is not a particularly new idea and has been done for years now by a wide range of game development companies. Already, 45 % of people who play mobile games play online multiplayer games at least once a month, according to research described in the SNAP Mobile literature on *www.forum.nokia.com*.[4] However it's become pretty clear that there exists a useful abstraction of multiplayer functionalities which are independent of any particular game – such as challenging someone (or anyone) to a new game, ranking players, or sending game packets using robust network protocols, friend lists, authentication, etc. This is what the Scalable Network Application Package (SNAP) from Nokia addresses.

Earlier in the book, in Chapter 5, Jo talked about the Arena technology developed for the Nokia N-Gage. The SNAP Mobile SDK is the Java ME client API for interfacing to the same backend framework, as shown in Figure 9.6. SNAP works for pretty much any Java ME device currently on the market. The only requirements are MIDP 2.0 and CLDC 1.0. It's included with the Sun Java Wireless Toolkit or you can download it separately from Forum Nokia (the URL is far too long to type in, but you can find it by searching for 'SNAP Mobile SDK') or from the SNAP Mobile website at *snapmobile.nokia.com*. It comes with a suite of sample applications to get you started straight away as well as an emulator to run locally to simulate the SNAP framework servers.

SNAP Mobile allows developers to focus on the game code rather than on network programming. Which makes a lot of sense – all you need to do is implement a few interfaces which serve as callbacks for the SNAP servers, and you've built your first SNAP Mobile game.

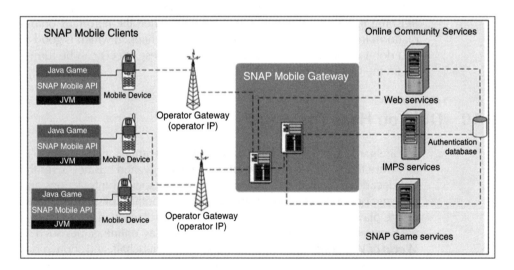

Figure 9.6   SNAP framework architecture

---

[4] Additional information, specifically for SNAP developers, can also be found on Forum Nokia at *www.forum.nokia.com/main/market_segments/games/snap_mobile.html*.

Clearly, this is going to speed up the development cycle and lower budgets immensely, since, generally, game programmers aren't necessarily network programmers.

The framework supplies standardized access to a variety of services from a SNAP game via a robust architecture. It provides a single point of contact between Java ME clients and community services, authentication, and account creation, session management, scalability, load balancing and routing services.

SNAP Mobile features fall naturally into three main categories:

- in-game community
- in-game connected game playing
- out-of-game community features.

9.10.1 In-Game Community

SNAP games allow users to access some pretty cool community features while still playing the game:

- friends list – the user's contact list in the online gaming community. This is a list of people you can chat with, send messages to, or challenge to a new game. You can also invite others to join and accept invitations to join other lists

- instant messaging/chat – send messages to friends and other users whether they are offline or online or whether they are in a lobby or a game room

- presence – indicators identifying friends who are online, offline, playing the same game or not to be disturbed. This has the effect of actually encouraging game playing

- unique user identities – SNAP supports online handles rather than using phone numbers or real names. This handle is retained by the user and used in all interactions with the SNAP framework – whether accessing the gaming community via a PC or a handset

- rankings – player statistics (scores and standings), top-ranked players and proximity rankings (ranking with respect to other players of similar standings) are all calculated as SNAP games report scores back to the SNAP servers for centralized storage. Again, this encourages online game playing and competition between players.

9.10.2 In-Game Connected Game Playing

SNAP Mobile has a variety of concepts that support multiplayer game playing:

- lobbies and game rooms – these are virtual concepts: games take place in game rooms and lobbies contains game rooms. This offers a place for users to meet, chat and play

- head-to-head connected game playing – users can play against one another from start to finish via peer-to-router-to-peer game sessions

- pervasive world games – always on, multiplayer sessions means that users from anywhere in the world can enter play and leave at any time

- matchmaking – SNAP Mobile has four modes for this: Challenge mode (where one player specifically requests a certain opponent), Random mode (automatically assigned to any available player in any game room), Join mode (player joins a specific game room), and Sort start mode (which uses a configurable load balancer to place users in game rooms).

9.10.3   Out-of-Game Community Features

SNAP Mobile offers a variety of concepts that support out-of-game community features:

- rankings – allow a more detailed view of the ranking statistics that are accessible from within a game as the interface isn't limited to the mobile screen

- support – phone settings, FAQs, trouble shooting guides, and so on

- featured game pages – game pages, information and screenshots from popular games

- message boards – for messaging regarding games and community-related topics. This is content-moderated and accessible from PCs only

- news and events – used to inform the community about new activities and upcoming attractions.

However you read it, the take-home message here is that the SNAP Mobile framework is designed to create and encourage a vibrant online community for multiplayer game playing. Nokia is working closely with network operators to ensure SNAP is supported by them, which makes a lot of sense, since it generates revenue through use of network bandwidth.

The SNAP SDK includes the mobile client API as a JAR file (sm-api.jar) as well as the framework emulator. You also get an API compatibility test MIDlet from Forum Nokia (***www.forum.nokia.com/main /market_segments/games/api_compatibility_test.html***) that you can use to help you to adhere to certain guidelines in order to get your game

SNAP certified which is a pre-requisite of using the SNAP framework servers. These involve mandatory requirements as well as best practices.

Every game has a globally unique class ID (GCID) which is centrally assigned by the SNAP Mobile team and must be hard-coded in your game source (you can't read it in from a JAD file as an attribute). It is used both to identify the game and as one of the arguments passed to the server during the login process. The API consists of the two packages com.nokia.sm.net, which contains classes used to communicate with a SNAP Mobile game server, and com.nokia.sm.util, which contains utility classes.

Server communications are done using an instance of the ServerComm class, and MIDlets register themselves with this communications channel by implementing the SnapEventListener interface and calling the addSnapEventListener method. This interface defines two methods which are used as asynchronous callbacks by the SNAP servers to notify the client of events.

```
public interface SnapEventListener{

  void processEvents(Vector list);
  void processServerError(int code, String message,
                                    int severity);
}
```

The methods of the ServerComms class are all pretty self-explanatory – examples include the acceptBuddyRequest, joinGameRoom, sendGamePacket and reportScores methods, which obviously do what they say. Most methods return an ItemList which is a container for objects that can be retrieved by name as well as a number of convenience methods for getting typed named data. The interface described above actually receives a Vector of ItemList instances, each of which has an integer identifier describing the type of event.

We've seen that the SNAP Mobile initiative provides a wide range of facilities to help developers create multiplayer online mobile games using Java ME very quickly and easily. It is expected that by 2010, over 30 % of mobile games will be connected games, so this is a great way for developers and operators to generate heaps of revenue. There's a lot more information about SNAP that we don't have space to cover, so for more information about the statistics quoted, complete code walkthroughs of developing SNAP games using Java ME, white papers, and certification guides, go to *www.forum.nokia.com/snapmobile*.

9.11 And Another Thing . . .

This chapter has talked (at length) about MIDP game development on Symbian OS. In particular, we talked about focused game design for the

MIDlet life cycle, and game pausing as core concepts. We also covered a lot of technicalities relating to why you should use Java ME for game development as well as heaps about the science of writing a MIDP 2.0 game using the example game ***Third Degree***. And we saw how Nokia's SNAP framework reduces time-to-market for online multiplayer mobile games and allows game developers to focus on what they're best at (all the fun bits).

However it's important to remember that most of the techniques covered here equally apply to Java ME game development as a whole not just on Symbian OS. Java ME has been the work horse of mobile game development for almost a decade and has a long history of successes in the market (yes – and some titles that imploded too), and there's no reason to think that this is going to change anytime soon.

Unless some other technology rises out of nowhere and suddenly becomes more ubiquitous than Java ME in the mobile space, there's just no competition – not with a billion Java enabled handsets out there. So there's no reason not to get into it now.

Hopefully this chapter has convinced you of the benefits of game development using Java ME (it has certainly convinced me anyway). I said earlier that it's the best cross-platform development technology in the universe. This *may* be an exaggeration, but until we make extra-solar contact, and they've done it better than us, I'll stand by that.

Symbian OS provides a world-leading best-of-breed CLDC/MIDP combination that enables experts and novices alike to dive head first into the exciting world of creating mobile games.

So what are you waiting for?

9.12   Further Reading

If you're looking for some further reading on Java ME development, here is a list of suggested reading:

General Game Development

- *Core Techniques and Algorithms in Game Programming*, Sanchez-Crespo Dalmau, New Riders, 2003
- *Physics for Game Developers*, Bourg, O'Reilly, 2002
- *AI for Game Developers*, Bourg and Seemann, O'Reilly, 2004

Mobile Game Development

- *OpenGL ES Game Development*, Astle and Durnil, Premier Press, 2004
- *Developing Scalable Series 40 Applications*, Yuan and Sharp, Nokia Press, 2005

Java/Java ME Development

- *Programming Java 2 Micro Edition on Symbian OS*, de Jode, Symbian Press, 2004

- *Java 2 Micro Edition Application Development*, Kroll and Haustein, SAMS, 2002

- *Programming Wireless Devices with the Java 2 Platform Micro Edition*, 2nd Edition, Riggs, Taivalsaari *et al.*, Addison Wesley, 2003

- *The Java Programming Language*, Fourth Edition, Arnold, Gosling and Holmes, 2006

10

Games In Japan

Sam Mason
(Mobile Intelligence)

10.1   Introduction

In this chapter, we examine mobile game development for the Japanese market. This market is unique in the Symbian ecosystem and presents new technical and social challenges for game development. We start with an outline of the reasons why this market is different, then explore the Java technology that can be used in developing games to run on phones in Japan, which is called DoJa. Moving on, we will discuss issues to be considered when porting a MIDP 2.0 game across to DoJa 2.5oe with some thoughts on the ***Third Degree*** game from Chapter 9, and we close with some remarks on using the Mascot plug-in for 3D graphics development in DoJa games.

10.2   The Japanese Market

We're all used to the large differential between Japan and the rest of the world when it comes to electronics and technology, so it should come as no surprise that the mobile phone market in Japan challenges most Western perceptions of device utility. As a consequence, if you're going to try and break into the Japanese game market, it's a good idea to have an idea of the playing field before kick-off.

There are almost 130 million people in Japan of whom 75% are mobile phone subscribers of some kind. Over half of these (so more than 50 million people) are subscribed to services provided by the telecommunications giant NTT DoCoMo[1] (the 'DoCoMo' comes from the phrase 'Do Communication over Mobiles').

---

[1] You can find out more about NTT DoCoMo at ***www.nttdocomo.com***.

Think about this – as you're reading this paragraph, millions of people in Japan are using their mobile phones to scan product bar codes and RFID tags, to pay for public transport, to exchange reviews on places to eat, to make reservations, to access their bank details, or to purchase goods while out shopping.

The youth demographic is heavily game-oriented and more than comfortable with cutting edge technologies. In addition, market penetration, consumer expectations and technology adoption rates are far above anything the rest of the world is used to. The pace of the market is phenomenal – the turnaround on software and hardware releases alone (more on this later) creates a unique platform for game development.

To celebrate the sale of the 20 millionth Symbian handset in Japan, the folks at Symbian put together a light hearted look at their uses of mobile phone technology (check out the 'Boo Hoo For You!' video at *www.youtube.com/watch?v=1xQVnny0LSg*). While this video is tongue-in-cheek, it's clear that to get into this market, your products must appeal to a very different group of consumers. So Japan is way ahead of us and already doing many of the things that we're still getting impressed about. I'm writing about it, and you're reading about it, but they've already done it.

To push the metaphor further, the referee in this game is NTT DoCoMo. They produce their own handsets and development platforms. They also released the world's first 3G network service, Freedom of Mobile Multimedia Access (FOMA), in 2001. It is based on their own W-CDMA technology that they created when other standards at the time were not in any usable state. To meet their specific requirements, NTT DoCoMo chose not to use the S60 or UIQ platforms and, instead, developed their own platform called the Mobile Oriented Applications Platform (MOAP).

NTT DoCoMo have complete control over available content and technologies, and they have created methodologies to ensure that things stay that way. As you'll see in this chapter, the business model drives the software design and development process down to an extremely low level.

FOMA handsets run Symbian OS v8.1b which gives them all the benefits of the full EKA2 kernel without the platform security model of Symbian OS v9.x. These devices don't need it because, as Chapter 1 described, they are 'closed,' which means that you cannot install after-market applications written in native C++ on them. In addition, there is no support for the Java ME MIDP versions 1.0 or 2.0.

The obvious consequence of this is that while these are Symbian OS smartphones, you can't just go ahead and write games for them using native C++ and a Symbian OS SDK, or using a MIDP (well you *can* – they just won't install on any FOMA devices, which isn't a whole lot of use when you're targeting the Japanese market).

One other thing we need to talk about here is the i-mode service, which is the mobile Internet platform originally created for NTT DoCoMo subscribers that can be accessed from anywhere in Japan. Over 5000 official websites are accessible via the devices' start menus (although unofficially there's a lot more than that). It's an 'always on' service, so there's no logging in or out. It offers extremely inexpensive access because charges are based on the volume of data sent and received rather than airtime. There are also plans that charge on volume up to a threshold after which all usage is free.

The i-mode service was so successful in Japan, that it expanded worldwide and the i-mode Alliance was formed, consisting of more than 5 million subscribers spread over roughly 16 countries including the UK, Australia, Germany, France, Israel and Russia. However it has been met with mixed success and, as Chapter 1 mentioned, in July 2007 the UK telecommunications giant, O_2, announced that they would discontinue support for it by 2009, as they only had 250,000 subscribers.[2]

So far, I've said a lot about what you *can't* do, and at this point you're possibly wondering how you actually *can* build games for the Japanese market. Good. As you'll see in the next section, you can create small Java applications called 'i-appli' that leverage the proprietary DoJa profile technology developed by NTT DoCoMo. This is the only way to create and distribute installable games, unless you use Flash Lite, which Nigel will talk about in the next chapter.

So, that's the market you'll be targeting:

- advanced devices with limited availability for hardware testing outside Japan

- FOMA/i-mode and MOAP restrictions from NTT DoCoMo

- no platform security because the phones are closed

- no Symbian C++ or MIDP development options available

- sub-optimal Javadocs

- a new profile called DoJa.

Added to this, you probably can't speak Japanese either! As they say, 'Boo Hoo For You.'

10.3 Enter DoJa!

DoJa, the Java development profile used in place of MIDP on i-mode devices, was conceived and created by – you guessed it – NTT DoCoMo. DoJa was created to provide a controlled application development

---

[2] More information can be found at ***www.iht.com/articles/2007/07/17/business/imode.php***.

environment for third party developers while ensuring that these applications ('i-appli') are secure and robust, and that the content providers' intellectual property (IP) is protected. DoJa also simultaneously generates revenue streams for the operator through its technical design.

Like MIDP, DoJa is dual layered and sits on top of a CLDC 1.0 (1.1 for more recent devices). It is built on the same basic premises that you would have seen in the previous chapter; so the following concepts should be familiar:

- a lack of floating point support (although this has been added into Japanese versions)

- a generic connection framework

- reduced `java.util` and `java.io` package functionalities

- no finalization

- byte code pre-verification.

Instead of a JAD file, DoJa supports the concept of a more advanced application descriptor file (ADF) which is completely standardized and integrated into the framework. In this part of the world, i-appli applications are managed by the Java Application Manager (JAM) which provides many of the same services as the AMS does in Symbian OS MIDP.

There are a number of versions of the DoJa profile, most of which are only available for use in Japan. For markets outside Japan, NTT DoCoMo produced two tailored API versions which are known as 1.5oe and 2.5oe (the 'oe' stands for Overseas Edition). These versions do not expose or support many of the advanced functionalities we've covered (such as bar codes, biometrics, etc.), because the required technical and social supporting infrastructure isn't available outside of Japan. In any case, Western consumers generally aren't as willing to pay for such expensive handsets anyway (including me).

Anyone familiar with the excellent Java Community Process (JCP), the JSRs, the reviews of JSRs, waiting for handsets that support them, and dealing with subsequent fragmentation issues will appreciate that one of the key benefits to NTT DoCoMo in developing a proprietary profile such as DoJa is that it can bypass the JCP altogether. Since DoJa is their profile, they can just get the in-house developers to build a new version whenever they decide they need it. Not surprisingly, there are at least seven versions of the profile for use in Japan, in addition to the two overseas APIs, whereas MIDP is only tentatively on its third version after more than twice as many years. And since NTT DoCoMo are providing the handsets as well as the development platform, they can avoid the sort of fragmentation problems that are caused by optional API support on target hardware that have been the bane of Java ME developers worldwide for years.

Newcomers to DoJa development will find a large number of technical restrictions on what you can do and how you can do it. This can be frustrating after the relative freedom of MIDP 2.0 with the suite of optional extras provided by the latest Symbian OS phones. We'll go into this in more detail later, but here's a taste of what I mean. In DoJa 2.5oe:

- you can't open sockets using anything other than OBEX or HTTP and, in the latter case, only to the server that the application was originally downloaded from

- data transfer rates on any given HTTP transaction are limited to 5 KB upload and 10 KB download

- the size of the application JAR file is usually limited to a mere 30 KB

- you only get up to 200 KB for local storage in a binary random access file called the Scratchpad (there is no Record Management System)

- there is no access to PIM data (e.g. calendar information)

- there are no Bluetooth APIs

- there is no multi-tasking – so only one application can be running at a time.

As a direct consequence of these restrictions, application architectures revolve around server-centric (and hence *bandwidth*-centric) storage. After all, if you need to store a recorded voice tag for a video you just took with the onboard camera, you're hardly likely to be able to store the resulting video and audio byte streams in a 100 KB file – so, chances are, you will need to develop a server-based remote storage model for your application.

As another example, multiplayer games need some way for the handsets to communicate with each other. While this can be done using infra-red, it's a bit frustrating when the communications fail because you didn't keep the devices aligned correctly throughout the game. Usually, the solution here is to use Bluetooth, but DoJa support for this did not arrive until version 5.0 (for Japan only), so as a result, developers have resorted to using server-based 'messaging' for real-time distributed game play.

This is a whole lot of data traffic, and it's not hard to see why this is such a successful business model for the operator. DoJa applications, by design, require a lot more use of bandwidth than either MIDP or native Symbian OS applications, and this is exactly what the end users get charged for.

Since it has extremely well-defined boundaries, the DoJa profile is very easy to learn and you can build advanced applications quite quickly. There are some excellent online tutorials (see the Mobile Developer Lab

website, ***www.mobiledevlab.com***, for more information), and a broad
developer support base both in Japan and Europe, found at ***www.doja-
developer.net/about***. However, there is one major drawback to the
technology – in order to keep the resulting JAR file below the 30 KB size
limit, DoJa developers generally have to adopt sub-optimal programming
techniques (actually they're appalling, but I'm trying to be nice).

To some people, this won't matter, but I'm really not a fan of this.
In general, the design patterns for DoJa applications, whether games or
otherwise, tend to consist of a main class that starts the application and
then another class that actually contains it (yes, you read it correctly – all
of it in one big class). My reaction to this is a sound "Umm . . . no."

Things are changing though – the latest versions of DoJa allow for more
flexibility, and many of the above restrictions either have been relaxed
or do not apply. At the time of writing, the current version is 5.0, which
allows up to 1 MB of memory to be shared between the JAR file and the
Scratchpad. It also supports GPS, RFID reading (FeliCa[3]), a gesture reader,
ToruCa (for transactions using dedicated read/write hardware), 3D sound,
matrix maths libraries for 3D graphics, and support for a number of effects
useful for game development, such as fog, lighting, transforms, textures,
and collision detection (in 3D) based on OpenGL ES 1.0.

There's only one catch (there's always one) – all this is designed for use
exclusively in Japan, and it's really hard to understand javadocs written
in Japanese (although arguably, most javadocs I've seen to date in *any*
language may as well have been in Japanese). As I mentioned above,
2.5oe is the latest DoJa API available for work outside Japan, so we're
going to have a look at this, and see what it *does* let you do.

10.4   DoJa 2.5 Overview

The easiest way to get started in DoJa development is to download the
DoJa 2.50e development kit from the NTT DoCoMo website at ***www.doja-
developer.net/downloads***. Make sure that you grab the development kit
user guide, release notes, and API docs while you're there. The installation
process is straightforward, and the only thing to note here is that you can
(and should) install the plug-in for Eclipse (at least version 2.1.1) via the
custom installation option. Figure 10.1 shows the installation in progress.

When you run the emulator, you get a start-up screen where you can
open or create new DoJa projects. The emulator does not include a text
editor, but you can set up an external editor of your choice via the Edit
menu. The default is Notepad. Of course, the better option here is to just
use Eclipse in the first place, but we'll get to that in a bit.

---

[3] ***www.nttdocomo.co.jp/english/service/imode/make/content/felica***.

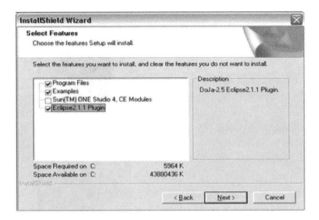

Figure 10.1 Installing the DoJa toolkit

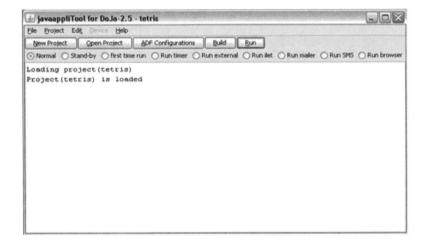

Figure 10.2 The javaappliTool interface

The javaappliTool (shown in Figure 10.2) allows you to load and run a project and includes an emulator supporting three device skins. They look a little strange, and probably the best one is shown in Figure 10.3.

I've already mentioned that the DoJa profile runs on top of a CLDC Java VM. In i-mode handsets, this is the KVM which can run with only 128 KB available memory. Well good – but you'll need a little more than that if you're going to build a useful GUI application, and the DoJa profile (also referred to as the i-mode extension API) enables you to do just that. The version 2.5oe profile consists of the packages shown in Table 10.1.

Traditionally, when you're learning a new technology, it's around about this point where the author walks you through the canonical **Hello World!** application. I find that approach rather irritating. With that in

Figure 10.3   DoJa device skin

Table 10.1   i-mode extension APIs

Package	Description
com.nttdocomo.lang	UnsupportedOperation and IllegalState exception classes
com.nttdocomo.io	Connection classes and exceptions for OBEX and HTTP
com.nttdocomo.util	Timers, events, encoders, and a phone utility class for making calls
com.nttdocomo.net	URL encoder and decoder
com.nttdocomo.system	Classes for sending SMS messages and adding new entries to the phonebook (although reading from the phonebook is unsupported)
com.nttdocomo.ui	Core widget classes for GUI development, sprite classes, and another phone utility class exposing vibration and backlight functionality
com.nttdocomo.device	Camera class allowing pictures and video to be taken using the onboard camera

mind, and given that this is a book on game development, let's fly in the face of tradition and build a quick little DoJa game called *Irritant*.

Start the javaappliTool via Start->Programs->javaappli Development Kit for DoJa-2.5->javaappliTool for DoJa-2.5, and click 'New Project' on the toolbar. When prompted, enter the project name and click 'Create' (shown in Figure 10.4).

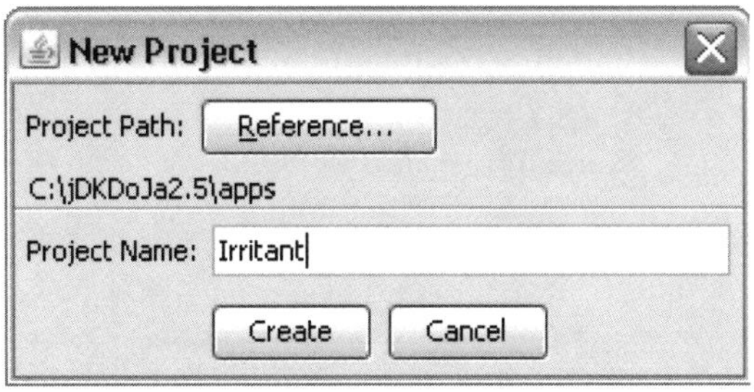

Figure 10.4 Creating a new DoJa project

Open Windows Explorer and navigate to the project's directory under your DoJa installation. On my machine, this is located at c:\jDKDoJa2.5 \apps. You should see a new directory named Irritant here, which holds the project files as described in Table 10.2.

Table 10.2 Project directory structure

Directory	Description
bin	Holds the ADF file (.jam) and JAR executable
res	Location for any image, video or audio resources to be distributed with the application
sp	Location for the Scratchpad file
src	Source files location

Next, choose Project->Make New Source File, call it Irritant and click 'Create'. This will create a basic source file in the src subdirectory of the project and open it for editing. When you've done that, enter in the code as shown below.

```
import com.nttdocomo.ui.*;
import com.nttdocomo.io.*;
import java.util.*;

public class Irritant extends IApplication implements SoftKeyListener{
  private final int BOUND_MIN        = 1;  // input bounds
```

```
private final int BOUND_MAX      = 10;
private final int BASICALLY      = 0;   // levels of irritation
private final int REALLY         = 1;

public void start(){
  MediaImage mi = MediaManager.
                              getImage("resource:///symbian.gif");
  try{mi.use();}
  catch(ConnectionException e){}
  Image image = mi.getImage();
  int level = BASICALLY;               // parse command line
  try{
    String[] args = getArgs();
    if(args[0] != null && args[0].equals("/i") && (args[1] != null)){
      level = Integer.parseInt(args[1]);
      if(level < BASICALLY || level > REALLY) level = BASICALLY;
    }
  }
  catch(Exception e){level = REALLY;}
  MainPanel mainPanel = new MainPanel(level, image);
  mainPanel.setSoftLabel(Frame.SOFT_KEY_1, "END");
  mainPanel.setSoftLabel(Frame.SOFT_KEY_2, "");
  mainPanel.setSoftKeyListener(this);
  Display.setCurrent(mainPanel);
}

public void softKeyPressed(int no){}     // from SoftKeyListener

public void softKeyReleased(int no){
  if(no == Frame.SOFT_KEY_1)
    IApplication.getCurrentApp().terminate();
}

class MainPanel extends Panel implements ComponentListener{
  private int iLevel;
  private Label iLabel = new Label("Enter a number from "
                              + BOUND_MIN + " to " + BOUND_MAX + ":");
  private TextBox iNumBox = new TextBox("",5,1,TextBox.DISPLAY_ANY);
  private Button iBtnOK = new Button("OK");
  private String appTitle = "Irritant v1.0";
  private Random random = new Random();

  public MainPanel(int irritationLevel, Image image){
    iLevel = irritationLevel;
    ImageLabel imageLabel = new ImageLabel(image);
    add(imageLabel);
    add(iLabel);
    add(iNumBox);
    add(iBtnOK);
    setComponentListener(this);
  }

  public void componentAction(Component source, int type, int param){
    if(source == iBtnOK){
      int entry = getEntry();
      if(entry != -1){
        int guess = Math.abs(random.nextInt()) % BOUND_MAX;
        int answer = ask("Was the number " + guess + "?");
```

```
          switch(iLevel){
            case BASICALLY:{
              showError((answer == Dialog.BUTTON_YES)
                ? "No it wasn't" : "Yes it was");
              break;
            }
            case REALLY:{
              showError((answer == Dialog.BUTTON_YES)
                ? "Fine - I win." : "You should have picked " + guess);
              IApplication.getCurrentApp().terminate();
                  break;
            }
          } // switch
          iNumBox.setText("");
        } //if
      } // if
    }

    private int getEntry(){
      int entry = -1;
      String attempt = null;
      try{
        attempt = iNumBox.getText().trim();
        int value = Integer.parseInt(attempt);
        if(value < BOUND_MIN || value > BOUND_MAX)
          throw new IllegalArgumentException();
        else
          entry = value;
      }
      catch(Exception e){
        showError("I said between " + BOUND_MIN + " and " + BOUND_MAX);
      }
      return entry;
    }

    private int ask(String question){
      Dialog dialog = new Dialog(Dialog.DIALOG_YESNO, appTitle);
      dialog.setText(question);
      return dialog.show();
    }

    private void showError(String message){
      Dialog dialog = new Dialog(Dialog.DIALOG_ERROR, appTitle);
      dialog.setText(message);
      dialog.show();
    }
  } // private class
} // class
```

Once the code has been entered, save and close the file, and select Build from the toolbar. If the build fails with the error message 'javac: target release 1.1 conflicts with default source release...,' it means that you are not compiling with 1.1 class file compatibility. As a work around, go to the Edit menu and select 'Use sun.tools.javac.Main.' You should now be able to compile the project without error. Start the game by clicking 'Run' on the toolbar.

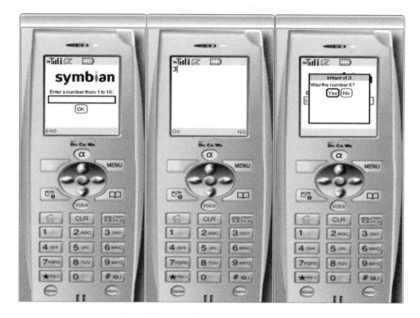

Figure 10.5   The *Irritant* demo application

The DoJa emulator, shown in Figure 10.5, can be a bit disconcerting when you're used to working with S60 or UIQ SDKs, and it takes some getting used to. To enter a number in the text box you need to select the text control first and then click it using the navigation keys. This then opens up an edit screen (akin to the Avkon Settings list default screen for text input) where you enter data and choose 'OK' from the softkeys. This populates the main screen and you can go on from there.

Irritant, while contrived (and irritating), actually demonstrates a number of useful techniques for DoJa application development:

- how to load and display a local resource such as an image

- how to read command line arguments from the ADF file (see below)

- how to terminate the application in response to an action

- how to set the softkeys (there are only two allowed in DoJa) and listen for any events fired by them

- use of basic widgets to build an event-driven GUI

- use of dialogs for user feedback and input.

Two other points – first, it's important for a game to deliver what it promises (usually inferred from the title) and second, *Irritant* also highlights how easy it can be to frustrate or annoy a user when designing a game (or indeed any software). Confrontational responses elicit emotional reactions and are always a bad idea in game development so are best avoided.

As mentioned above, you can set command line arguments via the ADF (although this is limited to 255 bytes only). To do this for **Irritant** click the **ADF Configuration** button or select the Project->ADF Configurations menu item. In the AppParam field (see Figure 10.6), type '/i 0' or '/i 1' to set the irritation level to 'Basic' or 'Really' respectively.

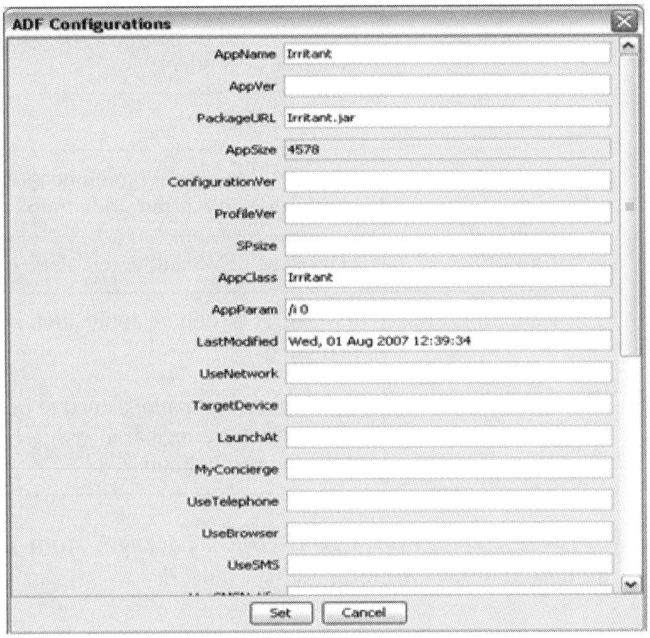

Figure 10.6 Editing the ADF with the DoJa toolkit

The ADF is where the metadata for the i-appli is defined and it is located in the project bin directory. For the example above, this file is created by the tool and is called `Irritant.jam`. This is analogous to the JAD file in Java ME applications. There are a large number of fields that can be set, most of which are optional.

Some of the more useful ADF parameters are defined in Table 10.3. Most of these exist for security reasons and the default behavior of the framework is to prevent application launch altogether if any fields contain invalid values. For a complete list of ADF parameters, refer to the development kit user guide.

Okay, that about covers as much of the high level side of the API as we need to. There's obviously a lot more, but it's more relevant to GUI application development for thick mobile clients than it is to game development. Anyone with any experience of Java AWT or the

Table 10.3 ADF parameters

Parameter	Description
PackageURL	The URL for the application binary. Communication by the application is only possible to the host and port defined here.
AppSize	JAR size in bytes. Set during compilation.
SPsize	Size of the Scratchpad in bytes. The maximum value here is device dependent.
AppParam	Start up parameters.
UseNetwork	Specifies if network communication is used. The only valid value here is the string 'http'.
UseSMS	Flags use of SMS. Either left blank or the string 'Yes'.
LaunchBySMS	If MSISDN is set here application launch via SMS is permitted.
LaunchAt	Used when applications need to be launched automatically at regular intervals. Format is I h where h is the interval in hours.

`javax.microedition.lcdui` packages from Java ME should have little problem working the rest out.

10.5   Porting MIDP to DoJa

In general, a game port from MIDP 2.0 to DoJa is pretty straightforward technically. Just how difficult this conversion process will be depends on the particular game and the functionalities it uses. A game that is largely dependent on CLDC classes as well as those in the low-level APIs only will be easier to port than one that leverages Bluetooth for multiplayer networking.

However it's all Java, and there is a significant amount of overlap between the low level APIs most relevant to game development, so working with the `Graphics`, `Canvas`, and `Sprite` classes is fairly easy. Double buffering is supported by default, but clipping, off-screen buffers, and partial drawing via regions isn't. Media support is different too: GIF and JPEG formats for images and, for audio, the Standard MIDI File (SMF) format is supported on all handsets and MFi format support is on most. It should also be apparent by now that any socket, datagram or serial communication mechanisms will need to either be removed or re-engineered.

Like MIDP 2.0, DoJa supports a mechanism that allows the developer to avoid polling the keys for user input in the game loop. In DoJa, the `Canvas.getKeypadState()` method performs the same function as the `GameCanvas.getKeypadStates()` method in MIDP 2.0 – in both cases, an integer is returned where each set bit represents a particular key stroke. Alternatively, the `processEvent()` method of the `Canvas` class can be overridden as shown in the code snippet below.

```
public void processEvent(int type, int param){
    if(type == Display.KEY_PRESSED_EVENT){
        switch(param){
            case Display.KEY_DOWN:{...break;}
            case Display.KEY_SOFT1:{...break;}
        }
    }
}
```

There are two types of timers in the DoJa API, and they operate differently from the MIDP equivalent. The `com.nttdocomo.util.Timer` class is used at an application level and notifies an observer class when its expiry event occurs. This type of timer remains active as long as the application does (or until it is explicitly cancelled). In contrast, the `com.nttdocomo.util.ShortTimer` class is only used with the low-level API, and events are sent to the `Canvas.processEvent()` method. In addition, this timer is automatically cancelled when the active screen is changed.

In summary, most MIDP application functionality can be ported to DoJa, sometimes requiring only a few source code changes. In fact, the most likely problem will be the requirement to keep the JAR executable below the 30 KB limit. This can require significant re-architecting of your MIDP application as the JAR size varies directly with the number of classes used and the number of methods in the classes, as well as the length of the identifiers used for methods, fields, and classes. A common way of reducing class size is to use an obfuscator which will replace readable names with shorter ones.

There are a number of other optimizations available to reduce executable size, including reducing the size of resources used, or moving access to them to a back-end server. Judicious use of string constants, using the same name for a method and a variable (so that the name reference can be shared), as well as reducing the number of methods and classes are all common techniques used to reduce class sizes and, therefore, the final JAR size.

As a demonstration, I've 'deflated' the **Irritant** demo application described above by applying some of these techniques (see the supplied source code **Irritant2**). The results are summarized in Table 10.4.

Table 10.4 Reducing JAR size (all sizes are shown in bytes)

Action	JAR size	Main class	Secondary class
Initial	4886	2183	3355
Reduce class names	4806	2113	3285
Reduce number of methods	4650	2113	3009
Shorten variable names	4564	2080	2899
Shorten string constants	4546	2074	2899
Change constant members to static	4496	1993	2899

Note that even in this simple application, it was possible to reduce the final size of the JAR file by almost 8 % (and, consequently, the readability of the code by almost 100 %, but the limitations of the technology dictate this approach).

This may not seem like much, but in a more complex game it could mean the difference between successful installation on a device and failure. Note also that these techniques could equally be applied to any MIDP games not simply DoJa ones; however, as I mentioned, Symbian OS does not impose JAR file restrictions for MIDP applications, so there's no excuse for doing so in a commercial environment.

Unfortunately, DoJa is an example of what happens when business strategy defines the functionalities and limitations of a technology. In order to overcome these, developers are forced to use poor programming techniques and, consequently, produce application code that is difficult to maintain or re-use. This has also been flagged as one of the reasons contributing to the poor acceptance of i-mode technologies outside of Japan, as section 10.2 described.

With this in mind, if we were to attempt a port of **Third Degree** (discussed in the previous chapter) across to DoJa, most of the changes we would need to make would revolve around the size of the resources (the 'Game Over' image alone has a transparent background and is 13 KB). These assets would need to be reduced in size (and probably quality) or moved to a server-based download approach.

As the game itself is fairly simple, and the logic and interactions are abstracted by the engine and controller classes, the core porting effort involves working around the fact that DoJa does not have the concept of layers or a `LayerManager` which are classes used extensively in most MIDP games (the **Third Degree** engine itself sub-classes `LayerMan-ager`). However DoJa does have a similar but less sophisticated concept called the `SpriteSet` which has methods that can be used for sprite collision detection.

There are differences in the sprite classes themselves as well. In MIDP, we can set a transform and define a frame sequence. In DoJa, you have to define and supply a linear transform yourself, and there is no built in support for frames sequences – although it isn't hard to do this yourself.

We'd also need to change the use of the graphics contexts and user input handling. In DoJa, the `lock` and `unlock` methods of the graphics class are used for double buffering (so there's no explicit calls like `flushGraphics`). While DoJa has a very similar mechanism to MIDP for getting the key state during the game loop, you may need to use the `processEvent` for handling canvas key presses in certain cases. Lastly, threads are generally shunned due to their overhead, in favor of a single application loop that blocks until the game ends; so the main game loop may need to be re-worked.

For all this, many i-mode devices have such limited memory constraints that it may not be possible to get acceptable performance regardless of these issues. In this case, you'd be left with no option other than simplifying the game world itself – reducing animations, including fewer objects in the world, simplifying the physics, etc. This should convince you that porting from MIDP to DoJa can often be more of an art than a science.

10.6   DoJa and Eclipse

I've talked a lot about how developers to date have worked around device constraints in DoJa games by placing resources on servers and pulling them down as required. Also, while the DoJa javaappliTool does the job of a basic development environment, it's hardly an IDE and wouldn't cut it in a commercial project – so let's address these two things and use Eclipse to access a server based resource for the **Irritant** demo we built earlier. In what follows, I've used Eclipse 3.2 with the DoJa plug-in and Tomcat as a local web server.

Start up Eclipse and choose File->New Project. With the plug-in installed, you should see the 'DoJa-2.5 Project' option under the Java node (see Figure 10.7). Choose this, and complete the wizard by clicking **Next** and entering your workspace location and project name (**Irritant3** in this case).

Before we go any further, we need to ensure 1.1 class file compatibility so right-click the project node and go to Properties. Choose the Java Compiler node and check the 'Enable project specific settings' option. Uncheck 'Use default compliance setting' and set the 'Generated class files compatibility' to 1.1. This will also require you to set the source compatibility to 1.3 (see Figure 10.8).

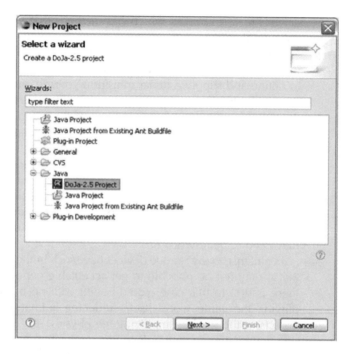

Figure 10.7   Creating a new DoJa project in Eclipse

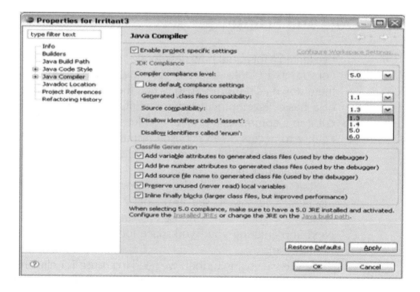

Figure 10.8   Setting class file compatibility

We're going to change **Irritant** to get the image file of the Symbian logo from our local web server. In this case, I simply placed a copy of the logo at the root of the default app on Tomcat, so the URL will be **localhost:8080/symbian.gif**.

Let's declare our intention to use network connections in the ADF by choosing Project->DoJa-2.5 Setup and entering the value 'http' in the UseNetwork field, as shown in Figure 10.9.

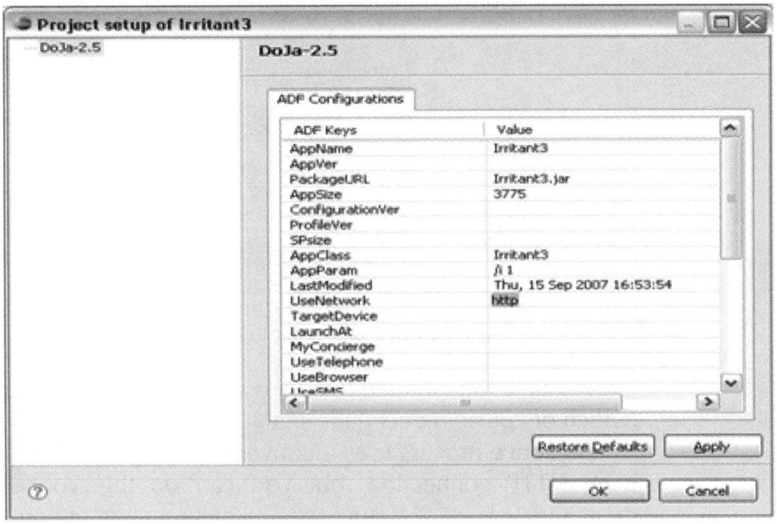

Figure 10.9 Setting network access in the ADF

Next, we change our network settings to define the source URL for the application (as represented by the ADF), by choosing Window->Preferences->DoJa-2.5 Environment, and enter the URL of your server in the ADF URL field as shown in Figure 10.10.

Okay, now comes the hard bit. Change the line that initializes the `MediaImage` from:

```
MediaImage mi = MediaManager.getImage("resource:///symbian.gif");
```

to:

```
MediaImage mi =
    MediaManager.getImage(IApplication.getCurrentApp().getSourceURL()
                                              + "/symbian.gif");
```

And that's it. Make sure your web server is running and that the logo is accessible at the location you've specified, and run the demo. You should see the application run as before, but remember that, this time,

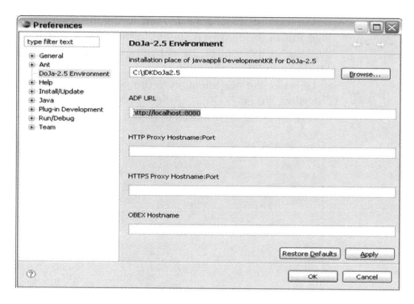

Figure 10.10   Setting the source location of the application

the resource was loaded from the network instead of from the JAR file (which of course reduced the size).

The above process uses a convenience method that wraps an underlying HTTP connection, but you can do this yourself utilizing the `com.nttdocomo.io.HttpConnection` class. If you do, then you're not limited to image files, but can pull down *any* type of data you wish (map or level data, 3D meshes, models, sound, etc). This is a powerful technique that allows DoJa to work around many of its resource limitations.

10.7   Mascot

The Mascot plug-in allows you to include high-performance 3D graphics in your DoJa games. It is supplied as a DLL which you download from ***www.mascotcapsule.com/toolkit/docomo/en/index.html*** under the 'Library' section (it's hard to see). Unzip it and place the DLL in the `bin` sub-directory of your DoJa toolkit and, woo-hoo, you're a 3D DoJa developer!

Three years before the Khronos group got together to think about mobile 3D graphics, HI Corporation released Mascot Capsule Engine Micro 3D Edition as the first commercially deployed embedded 3D rendering engine solution.[4] By 2003, more than 30 million handsets worldwide had a version of the Mascot engine in their firmware.

---

[4] ***www.mascotcapsule.com/en/***

Mascot is basically a set of native functionalities written in C with Java wrappers so that they can be used with Java games. In 2004, OpenGL ES was defined, so V4 of the Mascot engine includes both the V3 API and an implementation of the Mobile 3D Graphics API for J2ME (JSR 184). This uses OpenGL ES where hardware support is available, otherwise, it uses a pure software rendering implementation. It can run on almost any platform (Symbian OS, Java ME, BREW, Linux, Palm OS and Windows Mobile) and isn't dependent on any particular architecture, although V4 requires at least ARM-9 class handsets.

The DLL provides concrete implementations for the optional DoJa 2.5oe package `com.nttdocomo.opt.ui.j3d`, the contents of which are shown in Table 10.5 below.

Table 10.5 Core 3D graphics classes for DoJa 2.5oe

Class	Description
Figure	Represents a 3D data set for a model. Can be initialized from a file in MBAC format. Includes support for posture changes with bone animation and supports texturing.
ActionTable	Stores a series of posture changes for a figure.
Texture	Texture class that can also be used for environment mapping (called ambient mapping in Mascot terminology). Only supports uncompressed 8-bit BMP images. The class also supports both normal shading and toon shading models (toon shading is a technique that makes CG look hand drawn).
Vector3D	Represents vectors in 3D space with fixed point components (where 1.0 maps to 4096). Vector methods include cross product, normalize and dot product.
PrimitiveArray	Used for colour, vertex, point sprite, texture coordinate, and normal arrays.
AffineTransform	A 4 × 3 transformation matrix used for perspective and rotation (no translation or scaling). This is used to set the position of the camera in a scene via the Graphics3D context as well as manipulate world objects. All elements are fixed point numbers.
Math	Contains optimized fixed point implementations of the sin, cos, atan2 and sqrt function all commonly used in 3D calculations.
Graphics3D	This is the interface defining the graphics context functionality. It defines methods for lighting, shading, scaling and sphere mapping.

Whole libraries have been written on 3D graphics and it's outside the scope of this chapter to take this any further so we'll have to leave it there (although I'd love to talk about it some more). But 3D graphics work on mobiles is probably the most exciting (and growing) area in IT today, so have a look at the tutorials on the DoJa Developer Network and the HI Mascot site if you want to investigate this further.

10.8   Tokyo Titles

We were lucky enough to have Sam Cartwright, who's one of our reviewers, make it to the 2007 Tokyo Games Show just as we were putting this chapter together, and he's agreed to give us a bit of a taste of what we can expect in the Japanese market over the next 18 months. Here's what he had to say:

> ''At the 2007 Tokyo Game Show, NTT DoCoMo's booth featured a 'Mega Games Stage' that showed of some truly amazing DoJa 5.0 'mega games' running on their latest 904i series handsets.
>
> Many publishers were featuring games that immerse the player in full 3D environments created using the OpenGL-ES graphics library, a feature new to DoJa 5.0. Also on display were complete games or technology demonstrations featuring innovative uses for 3D sound, or the acceleration sensor and gesture reader that also make their debut in version 5.0.
>
> Without a doubt, the application that stole the show for me was Namco's new version of **Ridge Racers Mobile**. Full 3D graphics in WQVGA at 400 × 240 pixel resolution result in a stunningly beautiful game. At first glance you'd be forgiven if you could not tell the difference between the mobile and PSP versions.
>
> Another favorite was Hudson's mobile version of their Wii game **Kororinpa** (released as **Marble Mania** overseas). As with the Wii version, you must twist and turn your mobile handset to maneuver your marbles around various obstacles in the 3D game world. This is achieved using the AccelerationEventListener that allows the programmer to track the acceleration of the handset along the x, y and z axes, along with pitch, roll and screen orientation.
>
> Sega incorporated the gesture reader into their official **Beijing 2008** Mobile Phone Game. The gesture reader uses the phone's camera to track the player's movement (or gestures). In the 100 meter sprinting event, you must run on the spot in front of the phone's camera in order to move your game character along the track. The faster you run, the faster your game character runs.
>
> While the technical possibilities of the new DoCoMo line up is certainly impressive, many exhibitors still chose to focus on creating playable and enjoyable (read simple) games. While companies like Gameloft certainly haven't pushed the limit of the technology, they have delivered a line up of enjoyable games that can be played in short sessions without becoming frustrated at in-depth story lines or unusable controls. Gameloft's series

of best-selling 'sexy' games reminds us that in the end, playability, re-
playability and enjoy-ability is what sells games (and throwing in as much
adult content as the Japanese carriers will let you doesn't hurt either)."

10.9   At The End Of The Day

You know, it's actually pretty amazing that you can really write DoJa games so easily and then sell them in the biggest and most vibrant market on Earth. It used to be that you had to sign up as a developer or distributor for security reasons, but that's gone the way of the dodo. These days, i-mode users purchase and install DoJa games all the time, so it's a fantastic opportunity to get your product out there and working for you.

The only remaining issue is the difficulty in getting English-based development support. The DoJa Developer Network at ***www.doja-developer.net*** has some basic translated tutorials, but they tend to gloss over a number of deeper issues. There are plenty of websites devoted to DoJa development, but they're pretty much all in Japanese, so at this stage, I can't really recommend those to you (since I only know how to say 'hi').

By far, the best website for English-speaking developers outside Japan is the Mobile Developer Lab (MDL) at ***www.mobiledevlab.com***. The MDL site includes a number of articles, discussion forums and, most importantly, some translations of the API documentation for newer versions of the DoJa profile. This site was created and is personally maintained by Sam Cartwright, one of the top DoJa game development experts. Sam is based in Tokyo and has also reviewed, contributed to, and advised on the content of this chapter.

DoJa development is one of the more challenging game development technologies around and is also one of the most interesting. It sits astride the cutting edge of progress being constantly exposed to the very latest in mobile device hardware evolution. The great thing about that is that when a technology pushes boundaries, developers tend to get pushed too. When that happens, your exposure to innovation increases and consequently so does your professional skill set. And that's almost never a bad thing.

Hopefully, as time goes on, DoJa development will become more mainstream and take its place in the Symbian ecosystem alongside Symbian C++, P.I.P.S, Open C, MIDP and Flash Lite as a core mobile game development technology in the world outside of Japan.

But only time will tell, so we'll just have to wait and see, won't we?

11

Flash Lite Games on Symbian OS

Nigel Hietala

11.1 Introduction

Flash is all about great bitmap and vector graphics with animation. Flash abstracts away the complexity that C++ development normally involves, allowing the developer to just work directly on polishing the actual game. Developing with Flash can be productive, fast, rewarding and genuinely fun. Many developers are currently unaware of what Flash Lite is and how to get started developing with it. This chapter provides an introduction to Flash Lite and describes how it is particularly well suited to mobile games. I'll also discuss the powerful way that dynamic graphics and animation are abstracted via the Flash Lite runtime and API.

The chapter explores the available tools, which include not only Flash CS3 Professional, but also Adobe Creative Suite, and a range of third party and open source products that are necessary to make Flash development productive. Finally, I'll conclude with details of the Flash ecosystem, which is a combination of companies and communities that support, educate, and provide software extensions to Flash.

11.1.1 The Flash Platform

Flash may be a recent technology for mobile phones but it has already undergone a decade of development and use on the Internet. Flash started from humble beginnings as a tool and plug-in to enable vector animation for websites. Today, it has superseded its animation roots and is now, like Java, a sophisticated runtime.

Figure 11.1 shows the three main versions of the Flash platform on which you can develop. Flash Lite is the mobile incarnation of Flash. Version 1.0 was launched first in Europe and Japan in 2003. Version 1.1 followed shortly and was also made available in the US.

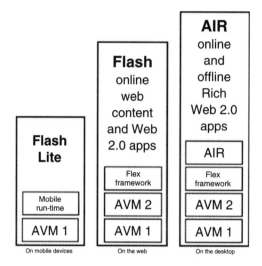

Figure 11.1   The Flash platform

Version 1.0 of Flash, for the web, was originally launched a decade ago.[1] Now at version 9, Flash Player penetration is at 90.3 % for Flash Player v9, 98.5 % for v8, and 99.3 % for v6 and v7.[2]

At the heart of Flash is the ActionScript Virtual Machine or AVM. Flash 9 provides a new virtual machine AVM 2. As Figure 1.1 shows, the original AVM that powers Flash Lite is still included in Flash 9, allowing backwards compatibility with all the older content available.

AVM 2 provides a huge performance jump through the introduction of a new version of Flash's scripting language called ActionScript 3.0 and a leading edge runtime technology featuring a just-in-time compiler. This same technology recently became a Mozilla open source project and will power the JavaScript side of a future version of the Firefox web browser.[3]

Web 2.0 is currently receiving a lot of positive press. Web 2.0 is the combination of a number of technologies to deliver a new generation of Internet sites and services. These sites are also known as rich Internet applications (RIAs) due to their ability to deliver an experience closer to the sophistication of a desktop application. One of the key technologies in this new world is Flash, and a specialized framework called Flex.[4] The Flex framework is built on top of Flash 9 and offers a number of APIs, components, and frameworks that make it easier to create Web 2.0 applications. It is discussed in more detail in section 11.4.4.

---

[1] *www.adobe.com/products/flash/special/flashanniversary*
[2] *www.adobe.com/products/player_census/flashplayer/version_penetration.html*
[3] *www.mozilla.org/projects/tamarin*
[4] Flex recently became an Open Source project as well: *www.adobe.com/aboutadobe/pressroom/pressreleases/200704/042607Flex.html*.

The rightmost box in Figure 11.1 shows the latest evolution of Flash. The Adobe Integrated Runtime (AIR) allows sophisticated desktop applications to be built with Flash and HTML technologies. Simply put, this means that a Flash Lite game developer is gaining a reusable knowledge and skill sets that can easily be used to target other versions of Flash. For example, it would be simple to provide versions of a Flash Lite game that are both viewable and playable in a desktop web browser when people visit the game's website using their PC.

11.2 Flash Lite on Mobile Phones

Flash, as Flash Lite, is available for a wide range of mobile devices including Nokia Series 40, Nokia S60 (2nd and 3rd Edition),[5] UIQ, BREW 2.x/3.x and Microsoft Windows Mobile 5. With over 300 million Flash enabled mobile devices and handsets shipped so far, Adobe expects more than one billion Flash-enabled devices to be available by 2010.

Eighty percent of the mobile devices shipping in Japan today can run Flash software. Handango, Iguana Mobile, and MOBIBASE, three of the world's leading content aggregators, have signed up to accept Flash Lite content, so opportunities are good for Flash Lite developers to distribute their games.

11.2.1 Flash Lite Versions

Flash Lite 1.x is based on Flash 4 for the web. Although graphically powerful, version 1.x is not nearly as powerful as Flash Lite 2.x. The limitations include missing primitive data types, such as arrays, and inclusion of an earlier version of ActionScript that is not object oriented.

Flash Lite 2.x is based on Flash 7 for the web and includes ActionScript 2.0. This is an object oriented scripting language with a rich API. Modern software development involves object oriented programming, sophisticated code editors, and testing frameworks. It requires teams of several people with workflow between developers and designers. Only Flash Lite 2.x supports this level of development and so is the focus of the remainder of the chapter. For those wanting to work on a Symbian platform, at the time of writing (September 2007), Flash Lite 2.x is only available on S60 3rd Edition smartphones.[6]

Just before we went to press, Adobe formally announced Flash Lite 3, which will be available first on NTT DoCoMo phones in Japan and globally on Nokia's phones. The new version includes support for

---

[5] More than 50 announced Nokia Series 40 and S60 device models support Flash Lite.

[6] You can find a list of mobile phones and the version of Flash Lite they support here: *www.adobe.com/mobile/supported_devices/handsets.html*.

Flash video (e.g., as found on YouTube), and Adobe says the aim is to more closely replicate the experience of Flash on the desktop. At the same time, Forum Nokia launched the 'Creative Pros' community at **www.forum.nokia.com/creativepros** to help desktop Flash developers port their content to mobile devices. The site provides tools and SDKs, plus technical and business support, to help creative professionals reach millions of mobile device users in the marketplace with applications, content, and services for Nokia's S60 and Series 40 platforms.

11.2.2   The Suitability of Flash on Mobile Phones

Flash Lite's web-based past is what makes it such a well-adapted mobile technology:

- the compiled Flash binary is optimized to be compact for a fast download. This fits with the limited storage space many phones still have, and also makes the binaries suitable for download to the phone over the phone network

- Flash supports both vector and bitmap graphics

- web browser plug-ins have limited access to the CPU so Flash does not have a requirement for a powerful CPU. The processors used by Symbian smartphones are relatively powerful, when compared to some of the feature phones that Flash Lite can also run on, and the fact that Flash does not require a large amount of processing horse power makes it suitable for mobile devices

- Flash Lite runs in a secure sandbox, meaning consumers and network operators can be confident that only a small subset of mobile phone and network functionality can also be accessed, and security cannot be compromised

- With some optimization, Flash 7 content can be reused for Flash Lite.

11.3   Flash Lite Overview

The following is a short overview of the main parts of Flash Lite and how they relate to mobile games, covering details of the graphics API, the scripting language, and how the two work together to make games easy to develop.

The compiled binary that makes up a Flash Lite game is known as a SWF file. Figure 11.2 shows a SWF, which is divided into two main parts. First come the library assets, which are the vector, bitmap and sound files. The other half is the compiled bytecode created from any ActionScript. The SWF is compressed, allowing it to be downloaded quickly when

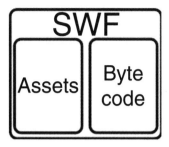

Figure 11.2 The parts of a SWF file

the user purchases a Flash Lite game. SWF files have to be opened with the Flash Lite player that is built onto the smartphone. Although this is acceptable for testing purposes, it forces the users to first find and open the Flash Player and then use it to open the SWF file, which becomes a barrier to use.

11.3.1 Symbian OS Specifics

A better way to deliver a game to a Symbian smartphone is by using a SIS file. This file is read by the Symbian software installer that allows the secure installation of any game or application. The SIS file contains a certificate guaranteeing the origin of the game, acquired from Symbian Signed, for which you can find more information on the Symbian Developer Network at ***developer.symbian.com***. Games provided as SIS files are visible in the application launcher and are shown with their own unique icon and name. This way, the user does not need to know which technology was used to create the game. They just have a quick and simple way to install and launch the game and to enjoy it. Packaging up SIS files requires the SDK for the phone. Alternatively, there are commercial tools[7] available that will take the SWF, an application icon, and create a SIS file.

Flash Lite applications are able to launch native applications. For example, a game could launch the web browser and have it load the website with more details about the game itself or other games that are available. Because Symbian OS is an open platform there are already third-party frameworks[8] that allow Flash Lite content to access additional device functionality. This includes phone book, messaging, camera, GPS, and Bluetooth access. Flash Lite can additionally play native content such

---

[7] SWF2Go is a professional toolkit for Adobe Flash Lite, which allows developers to create Symbian SIS installer and SWF launcher icon for Flash Lite applications. More information can be found at ***www.orison.biz/products/swf2go***.

[8] A good example is 'Kuneri Lite' found at ***www.kuneri.net*** or 'Flyer' (requires Python for S60) ***www.flyerframework.org***.

as video files, and it is simple to discover phone specific details such as the exact screen resolution.

11.3.2   ActionScript

ActionScript is the scripting language that makes Flash a runtime. It is based on the same ECMAScript[9] standard as JavaScript. Its syntax is easy to learn, and developers of Java and JavaScript will find it a simple step to pick up. ActionScript 2.0 is the version used for Flash Lite 2.x development. Memory management, as with Java, is handled by a garbage collection mechanism. There is no risk of broken pointers, memory leaks, and data type errors.

Unlike Java, ActionScript does not force everything to be written as a class. For example, 'Hello World' in ActionScript is simply:

```
trace("Hello World");
```

A real world example that works outside of a test environment would be:

```
var textBox:TextField; // Declare and type the textbox variable

// Add a TextField called textbox to the display
createTextField("textBox", 1, 0, 0, 100, 15);
textBox.text = "Hello World"; // Set the text to show Hello World
```

ActionScript is optionally typed. Typing variables and functions allows the compiler to give early warnings of problems. This flexible typing also extends to arrays, which can hold any type of object without the need for casting. Flash Lite has a flexible and forgiving language compiler. While great for beginners learning to code, this and other features, such as lack of runtime error messages, can make it harder to finalize a Flash Lite game.

11.3.3   The Benefits of OOP

ActionScript 2.0 offers an explicit, but optional object oriented programming (OOP) syntax. To ensure that a Flash Lite game reaches a commercial standard, some early architectural decisions are needed. To manage the code effectively, OOP must be used, and code split into separate classes. The benefits of Flash OOP are manifold, for example:

- the classes created can be reused

- multiple coders can work on the game in parallel

---

[9] See **www.ecma-international.org/publications/standards/Ecma-262.htm**.

- OOP techniques and knowledge such as design patterns can be used

- a range of advanced development tools are available.

Stricter and faster compilers, unit test frameworks, and entire Action-Script IDEs are only available for ActionScript that is split into OOP classes.

There is an alternative way of working that is a common Flash development style used by many designers. Small scripts can be placed on different parts of the timeline and are activated when the play head reaches them (see the MovieClip explanation below). It is possible to create an entire game on the timeline with the Flash CS3 Professional application.

However, a word of warning: although visually impressive games can be created in this way, having a large timeline, with assets all over the place, hundreds of separate frames, and code literally anywhere comes at a cost. Maintenance at the end of the project can be slow and expensive. It is laborious to navigate a project and move hundreds of objects around the timeline, and this approach generally means that only a single person can work on the game. It is initially a fast way to create a game but should be limited, in most cases, to only being used for aspects that require a quick prototype.

11.3.4 The MovieClip class

Within a game, MovieClip classes are used for animation, sprites, buttons, playing cards, UFOs, and explosions. Each individual MovieClip has its own timeline and play head. Figure 11.3 shows a MovieClip as well as a few useful properties, methods, and events that the class supports. The combination of properties, methods, and events that MovieClip supports allows an almost unlimited array of options for designing various parts of the game.

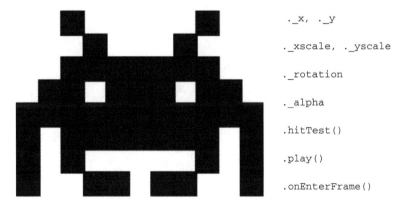

```
._x, ._y

._xscale, ._yscale

._rotation

._alpha

.hitTest()

.play()

.onEnterFrame()
```

Figure 11.3   A MovieClip and the useful properties, methods and events the class supports

As I mentioned, every MovieClip has its own timeline. The timeline is visualized in the Flash IDE and is similar to a filmstrip with separate frames. A powerful feature of the timeline is that the sound can be synchronized to match up with specific frames of the animation on a timeline. This can be used to enable lip-syncing of characters, or to ensure that the sound of a monster exploding occurs at the same time as the visuals. The play head relates to the current frame of the MovieClip that is being played at any given moment. The play head of any MovieClip can also be programmatically stopped or moved to a new position.

The independent timeline of a MovieClip enables some clever game functionality. The different states of a sprite can be placed on different frames of a MovieClip. For example frame 1 could have a laser cannon in its normal state, frame 2 could have it firing and frame 3 could have the graphic of the laser cannon exploding when an alien weapon has hit it. ActionScript can then be used to move the play head to the correct frame to show the correct graphic, or set to play a range of frames that show an animation. Laser cannons can then explode properly in a shower of pieces, not go bang in one single frame.

Using the Flash CS3 Professional tool, MovieClips can be stored as assets in the SWFs library. Specific ActionScript classes can then be linked to these MovieClips so that, when they are used at run time in the game, the linked class that controls them is automatically instantiated as well.

11.3.5   Layers and Child MovieClips

One way of working is to have all MovieClips on independent layers, something that is shown in Figure 11.4. This way of working may become progressively complex, but great versatility comes from adding child MovieClips to a parent. For example, if we had a MovieClip called 'moonBase,' we could attach several child clips to it, based on assets in the SWF's library.

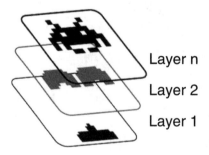

Figure 11.4   Independent MovieClip layers

```
var laserCannonClip:MovieClip;
laserCannonClip = moonBase.attachMovie("laserCannonAsset", "laserCannon1",
            moonBase.getNextHighestDepth());
```

We now have a laser cannon attached to the MovieClip of the moonBase. The enemy hits it with a shrink ray that also makes everything slightly transparent. In Flash, child MovieClips inherit the properties of the parent.

```
moonBase._xscale = 50;
moonBase._yscale = 50;
moonBase._alpha = 50;
```

This would shrink both the moon base and laser cannon as well as set their alpha transparency to 50 %. This allows sweeping changes to be applied to many MovieClips quickly and easily with small amounts of code. This way of working is useful with masks as all the child MovieClips will show through the mask that the parent clip is seen through. This saves the need for multiple masks to show related content.

A powerful demonstration of how ActionScript, MovieClips, and the timeline events combine is in how easy it is to create movement. As the timeline ticks from frame to frame, it generates an event called onEnterFrame.

```
moonBase.onEnterFrame = function() {
this._x++;
this._rotation++;
}
```

This would send the MovieClip spinning clockwise and traveling rightwards at however many frames per second the Flash game is set to.

There are entire books dedicated to ActionScript, some of which are animation and game specific. See the links provided in section 11.5 for more information about the multitude of books, videos, and websites that can fill in all the details.

11.4 Flash Lite Development Tools

11.4.1 Flash CS3 Professional

Flash CS3 Professional (the CS3 stands for Creative Suite 3) is the main tool for Flash development. Apart from a tool to package the binaries into a SIS file for installation onto the Symbian smartphone, for example, SWF2Go described in section 11.3.1, everything needed to develop a Flash Lite game is installed along with the Flash CS3 Professional (which I'll call simply the 'Flash IDE' or 'CS3' in the rest of this text). There is no need to download and install SDKs or scripting languages, or to set up environment variables. CS3 is available on its own or as part of a variety of suites that include other products in the Creative Suite including Fireworks, Illustrator and Photoshop.

CS3 is the first Flash IDE released since Adobe and Macromedia merged. It is a significant upgrade from previous versions and offers a new improved UI that, for the first time, is consistent with other CS3 applications, including Photoshop and Illustrator. A screenshot is shown in Figure 11.5.

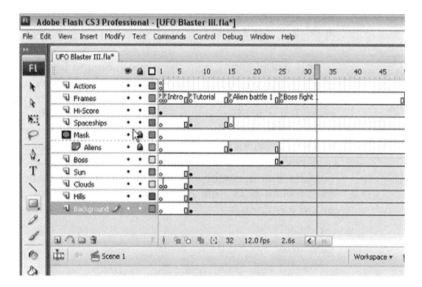

Figure 11.5   Adobe Flash CS3 Professional

The Flash IDE is a powerful tool for designers. It allows a library of graphics and sounds to be managed. It offers a range of tools for manipulating bitmap and vector graphics. For the design and art side of a game, the Flash IDE is a critical tool. Most of the time will be spent managing items on the stage (the electronic canvas that represents what will be seen on screen). The other key part is the visual timeline. This allows designers with little or no software development knowledge to create rich, dynamic, interactive game elements. As I described above, this flexibility can be used positively to try out ideas, but it can also be misused to create games that cannot be maintained. The timeline allows ideas to be quickly tried out. However, these elements should be re-factored to ensure the code and graphics are separate from each other.

On the software side, the Flash IDE provides ways to profile the SWF to allow optimization of the file size. For bitmaps, compression settings can be tweaked on a global level or for individual files. An ActionScript language reference specifically tailored for Flash Lite can be searched and browsed. For more sophisticated code editing, see the later tools section.

11.4.2 Adobe Bridge and Device Central

The Adobe Bridge is a separate component that ships with all applications that are part of the Adobe Creative Suite. It is dedicated to the management of assets such as JPEGs and PNGs. Even SWF files can be interactively previewed inside the Bridge. This is an invaluable feature in the late stages of development, when there may be many versions of the same game to choose from.

Something new with CS3 is Adobe Device Central, which is a dedicated mobile simulation environment, a screenshot of which is shown in Figure 11.6. For Flash, it speeds up application development by allowing games to be previewed and tested without having access to real phone hardware. Photoshop, Illustrator, and several other CS3 applications also use it to preview how graphics will look on a mobile phone. It speeds up the development cycle, as more work can be done on the desktop, and avoids the break in the flow that can occur when you need to stop to check how the game runs on a real Symbian smartphone.

Device Central is regularly updated with the device profiles for the latest phones that support Flash Lite. Other functions include the ability to simulate the speed of a mobile phone to ensure there are no performance surprises when the game is run on real hardware. Device Central is

Figure 11.6   Adobe Device Central CS3

even more valuable now that there are a growing number of phones for games developers to target and purchasing every compatible device is impractical.

11.4.3   Adobe Creative Suite

The full creative suite contains tools for areas such as video, sound and visual effects as well as web and print design. Many of these have nothing to do with games development, but three applications that do are Fireworks, Photoshop and Illustrator CS3.

Photoshop can be used for editing bitmaps and Illustrator can be used for vector graphics. Both tools have functionality that goes far beyond what Flash offers, and are critical for the creation of top grade graphics. Effects such as glow, drop shadows, and blur cannot be achieved in real time with Flash Lite. However, they can be applied to graphics by using Photoshop.

Fireworks combines many of the key bitmap and vector tools from both Photoshop and Illustrator, but in an easier to use, more compact application.

Now that Flash, Fireworks, Photoshop and Illustrator are part of the same suite and have similar interfaces and some level of integration, it is possible to ensure fast workflows between designers and developers.

CS3 applications are also extendable via plug-ins and scripts. Plug-ins provide new functionality, while scripts allow the automation of repetitive tasks. It is this ability to extend and automate things that means professional users of CS3 applications can free up more of their time to polish their games.

11.4.4   Eclipse IDE

The Flex framework builds on Flash 9. Adobe offers a specialized tool for building Flex and ActionScript 3.0 projects through a software development focused IDE called Flex Builder. Flex Builder is built on a free open source platform called Eclipse (more information is available at **www.eclipse.org**). It is also possible to use Flash Lite and ActionScript 2.0 in Eclipse. There is an open source project called ASDT, found at **osflash.org/projects/asdt**, that provides support, but its features are rather limited.

ActionScript developers should also have a look at FDT[10] (Flash development tool). It is a commercial plug-in for Eclipse and provides everything a developer would want. Eclipse itself brings source control integration, ANT builds, source file searching, and many other features. For the small overhead of having to write ActionScript class files, FDT

---

[10] **fdt.powerflasher.com**

offers class browsing, easy source file navigation, syntax highlighting, auto code completion, integrated Flash documentation, quick fixes, and real-time source file parsing. This real-time parsing results in seeing errors in the code as you type. Thus, instead of having to compile your code to see common mistakes, FDT shows them in real time. It can also provide warnings that allow better code styles to be enforced. It can, for example, warn of functions that have no return types. FDT allows compilation with both the Flash IDE and MTASC, which we'll talk about next.

11.4.5 MTASC

The Motion Twin ActionScript Compiler, or MTASC[11] as it is known, is a fast compiler that can noticeably speed up game development. It is an open source project and binaries are available for Microsoft Windows, Mac OS X and Linux. There are some important caveats that must be considered before use:

- it does not support all the code styles that Flash CS3 Professional will compile. So code from existing projects will need some porting

- it is possible to write code that works with MTASC that does not behave identically when compiled with Flash CS3 Professional.

For projects starting from scratch and with no requirement to later compile with the Flash IDE, MTASC speeds up development in the following ways.

First of all, MTASC only compiles ActionScript. A SWF full of assets will still need to be provided. MTASC will then compile the ActionScript into byte code and 'inject' it into the SWF to make a complete binary. The first speed gain is from not having to recompile the assets into the SWF, something that can take minutes, if there are lots of graphical assets and sound files to reconvert.

The second gain is from the speed in compiling. On a modern desktop over 100 classes can be compiled in under five seconds. This takes the agile development model of quick iterations, write some code and then build, to its full extreme. It allows the developer to write a couple of lines of code and then recompile the entire game in a few seconds to see the result.

The MTASC compiler also can be set to provide more verbose compilation details, as well as a strict mode that will refuse to compile code if it is not strictly typed. MTASC also includes a few other bonuses. Java developers may feel more at home by utilizing comment based metadata that can be added to facilitate typed arrays.

---

[11] *www.mtasc.org*

11.4.6 Xray

Flash Lite games can be tricky to debug. There are no sophisticated debuggers as there are with the other languages on Symbian OS, such as C++ and Java. Flash has a trace statement that can output log style messages to help find problems. However the use of trace is limited. Further debugging techniques, such as being able to recursively trace the contents of arrays and other objects, are needed. If you are used to other development languages you may expect proper logging with debug, warning and error messages. Xray[12] is an open source visual debugger that provides for your needs.

Figure 11.7 shows Xray, an application that is written entirely in Flash. Xray provides all the recursive trace and logging requirements a game developer needs. What Xray also does is that it takes advantage of the fact that Flash is a dynamic runtime. When the game is in any state, Xray can take a snapshot that results in a visual 3D tree, showing the entire hierarchy of all the visual (and some non-visual) elements.

Figure 11.7   A typical view in the Xray tool

Xray shows all the MovieClips and is an invaluable way of hunting down unneeded assets and speeding up a game. It becomes very easy to find problems, such as MovieClips, that have been attached off-screen.

---

[12] More information about Xray can be found at **www.osflash.org/xray**.

Xray then allows the editing of the properties of the MovieClips while the game is still running.

This can save a lot of time: MovieClips can be repositioned into exactly the correct pixel or set to the perfect level of alpha without the normal trial and error of changing code, recompiling, and then seeing if it works. Xray can be used to move MovieClips into correct positions and then the new property can be used in the final game code. Xray also plays a critical role in optimization. The more graphical assets on the screen, the slower the game is. Graphics can easily be hidden behind other graphics.

11.5 Flash Lite Ecosystem

The Flash ecosystem is the final key part of what makes Flash development so great. Flash may be powerful, but who is going to educate you to all its possibilities? What happens when support is needed with fixing bugs or selling a game? Who is going to keep you up to date on the latest Flash news and give you a place to mix with like-minded Flash fans? What about other tools, such as unit testing frameworks, animation frameworks, and many other code libraries? This is where the Flash ecosystem steps in.

Flash Lite game developers should start at the official Adobe Flash Lite developer site at **www.adobe.com/devnet/flash**. This site is the first place to go for important development information, documentation, and tutorials. The content developer kits (CDKs) can be downloaded from the site, along with profiles of new Flash-enabled phones for Adobe Device Central. The site also hosts the official Flash user forums, which offer a Flash Lite exchange, where games – both commercial and free – can be downloaded.

A variety of schools, colleges and private companies provide training in Flash, ActionScript, and other related needs. Direct access to the world's best trainers is available through electronic training videos that are available as DVDs or online. Two companies that provide these services are TotalTraining (**www.totaltraining.com**) and Lynda.com (**www.lynda.com**).

All the big technical publishers provide a multitude of books that cover literally every specific Flash need. Two publishers that specialize in Flash books are O'Reilly Media (**www.oreilly.com**) and FriendsOfEd (**www.friendsofed.com**).

Flash has a number of communities that have sprung up offering education, support, and extensions to the main Adobe offering. One community that should not be missed is the amazing open source community. Details of all the big open source projects, mailing lists, and other community offerings can be found at **osflash.org** and at **www.riaforge.org**.

References and Resources

Online Resources

Symbian Developer Network

Symbian Developer Network (***developer.symbian.com***) is a good resource for accessing the technical and commercial information and resources you need to succeed in the wireless space. Symbian Developer Network provides technical information and news, and gives you access to the tools and documentation you need for developing on Symbian OS platforms. You can also find discussion and feedback forums, example code and tutorials, a searchable list of FAQs and a developer wiki.

The wiki page for this book can be found at ***developer.symbian.com/ wiki/display/academy/Games+on+Symbian+OS***.

We'll post additional links that are of interest to game developers working on Symbian OS on the wiki site – do stop by and add your own favorites to the wiki page. We'll also use the wiki page to keep an errata list for this book, and to provide additional sample code, news and useful information.

The main Symbian Press site for this book, from which you can download a full set of sample code, is ***developer.symbian.com/gamesbook***. You can find the ***Roids*** game example code at ***developer.symbian.com/ roidsgame***.

Other websites of interest for those developing mobile games on Symbian OS include those described in the following section. Please note that the layout of external websites may change over time, and we will post updates to this book's wiki page when we find that the links specified below are no longer active.

Symbian OS Developer Sites

Forum Nokia: (*www.forum.nokia.com*)
The SDK for working on the S60 platform can be downloaded from Forum Nokia, as can the Carbide.c++ IDE, and other tools and utilities. Forum Nokia also hosts an extensive developer wiki and discussion forums, and provides a number of documents on general S60 development, as well as a games-specific documentation section which can be found at *www.forum.nokia.com/main/resources/documentation/games*. A discussion forum about creating games on S60 can be found at *discussion.forum.nokia.com/forum.*

UIQ Developer Community: (*developer.uiq.com*)
The UIQ Developer Community site has many resources invaluable to developers creating applications, content and services for UIQ 3 smartphones.

About Symbian, Symbian OS and UI Platforms

Company overview: *www.symbian.com/about/overview/hi/hi.html*
Fast facts: *www.symbian.com/about/fastfacts/fastfacts.html*

MOAP user interface for the FOMA 3G network: *www.nttdocomo.com*
S60: *www.s60.com*
UIQ: *www.uiq.com*

Standards

The Khronos Group: *www.khronos.org*

DoJa Development in Japan

NTT DoCoMo: *www.nttdocomo.com*
The Mobile Developer Lab: *mobiledevlab.com*
DoJa Developer Network: *www.doja-developer.net/about*

Game Sites from Nokia

N-Gage: *www.n-gage.com*
SNAP Mobile: *snapmobile.nokia.com*

Background Information about Mobile Game Development

International Game Developers Association, Mobile Game Development Special Interest Group: *www.igda.org/mobile*
Robert Tercek: *www.roberttercek.com*

The Authors

Mobile Intelligence: *www.mobileintelligence.com.au*
Mobile Radicals: *www.mobileradicals.com*
TwmDesign: *www.twmdesign.co.uk*

Books

By Symbian Press

If you are new to C++ development on Symbian OS, and wish to find out more about the idioms, toolchain and application framework, we recommend these books to get you started:

Developing Software for Symbian OS 2nd Edition, Babin. John Wiley & Sons.
Symbian OS C++ for Mobile Phones, Volume 3, Harrison, Shackman *et al.* John Wiley & Sons.

Other books in the Symbian Press series:

S60 Programming – A Tutorial Guide, Coulton *et al.* John Wiley & Sons.
Programming Java 2 Micro Edition on Symbian OS, de Jode, John Wiley & Sons.
Symbian OS Platform Security, Heath *et al.* John Wiley & Sons.

The following book is a more detailed reference book for creating C++ applications on S60 3rd Edition and UIQ 3:
Symbian OS C++ Architecture, Morris. John Wiley & Sons.

Symbian OS Internals, Sales *et al.* John Wiley & Sons.
Symbian OS Explained, Stichbury. John Wiley & Sons.
The Accredited Symbian Developer Primer, Stichbury and Jacobs. John Wiley & Sons.
Smartphone Operating System Concepts with Symbian OS, Jipping. John Wiley & Sons.

By Other Publishers

Small Memory Software: Patterns for Systems with Limited Memory, Noble and Weir. Addison-Wesley Professional.
Mobile Phone Programming: and its Application to Wireless Networking, Fitzek and Reichert. Springer Verlag.
The Business and Culture of Digital Games: Gamework and Gameplay, Kerr. Sage Publications Ltd.
Cross-Platform Game Programming, Goodwin. Charles River Media.
The Ultimate History of Video Games, Kent. Three Rivers Press.

Appendix: Airplay

Ideaworks3D

Native operating systems offer mobile game developers huge opportunities for revenues through richer, more interactive game experiences. There are currently considerable business challenges which may limit the ability to exploit these opportunities for games deployed over multiple platforms. Airplay SDK solves these problems through unique technology which has already been proven in many award-winning and best-selling mobile games, providing enormous cost savings and opening up new channels to the consumer.

What is Airplay SDK?

Airplay SDK from Ideaworks3D consists of a comprehensive C/C++ SDK and a set of tools for native mobile game development.

The solution comprises two environments: a development environment and a deployment environment. Both environments include a set of powerful and extensible tools and technologies within a framework of processes and workflow best practices. This combination of technology and know-how delivers significant cost savings during both development and deployment and enables a faster time to market.

The Airplay solution is the unique result of a six-year symbiotic relationship between Ideaworks3D's technology and game studio divisions. It is through building many of the most innovative mobile games in the world that our solution has been fine-tuned and thoroughly battle-tested.

Airplay consists of three components:

Airplay System – This is an easy-to-learn abstraction layer that is designed to relieve deployment headaches and offer a zero-effort porting solution by using a single platform-independent native game binary.

Airplay Studio – This is a set of fully featured OS-independent libraries and art tools, providing a highly-optimized, scalable art content pipeline and runtime graphics engine. The graphics engine can render using either the included software renderer – widely recognized to be the highest-performing in its class – or by taking advantage of any available hardware acceleration.

Using Airplay Studio tools, developers can rapidly produce console-quality mobile and handheld titles, and publishers can easily deploy each title across a variety of platforms, virtually eradicating the cost of porting QA, which results in a much faster time to market. The tools have been designed to exploit synergies across the entire console/handheld/mobile spectrum, and to allow the re-use of a variety of assets from console and handheld platforms (such as textures, geometry, video, speech, music, sound effects and even code). The tools also simplify porting from first-generation (e.g., PSone) consoles or PC titles. Each of the Studio components have been tested in-house by the Ideaworks3D Studio, which has used them to build games that have contained well over 1 MB of compiled code and in excess of 50 MB of game assets.

Airplay Online – As Chapter 5 described, Airplay Online is a flexible and powerful framework for online games. Airplay Online provides standard services for downloading levels and other assets, high score uploading and table downloading, identity management, profile management, and real-time multiplayer gameplay. In addition, it provides a framework that makes it easy to enable custom services to be written, for instance, implementing game logic on the server to make things simpler for the client.

Business Challenges in the Mobile Games Industry

The emergence of addressable native operating systems (Symbian OS, BREW, Windows Mobile, Mobile Linux, and potentially RTOS environments like Nucleus) has brought a new opportunity for games developers. The ability to run at native execution speeds opens up a new class of rich, interactive game experiences.

Native mobile game development currently comes at a price. Richer game experiences demand more complex development, often over longer periods and with a more skilled workforce. There is also an onerous porting headache between native operating systems. Entirely new platforms (e.g., Android) also appear periodically, and supporting these requires another large investment from the developer and publisher.

Even on a single platform, the rapid production of new handsets means that games developed and tested on a specific range of handsets miss new deployment opportunities only a short time after release. Deploying an existing game to a new handset often requires re-convening the development team, re-compiling the game code, testing and tweaking on the new handset, before uploading the new game binary to the operator. This is a costly and time-consuming exercise.

Airplay SDK recognises the massive opportunity for rich game experiences on native platforms, and addresses these business challenges as described below.

How Airplay Solves the Business Challenges

Airplay SDK solves these business challenges in the following ways:

1. **It reduces porting costs** almost to zero by providing a single game binary which runs across all native operating systems.

2. **It reduces code development costs** by allowing the entire mobile application to be developed on the desktop.

3. **It reduces code debugging costs** by providing full source-level debugging of the native ARM binary on the desktop.

4. **It reduces art development costs** by providing a 'scalable content pipeline' that supports hardware-accelerated OpenGL ES 1.1 as well as a best-in-class software renderer.

5. **It reduces the cost of deploying to new handsets after release** by separating the development environment from the deployment environment. A single game binary can be repackaged with an appropriate data set post-production and a SKU delivered to a formerly unknown device, without requiring the involvement of the original game development team.

Let's look in more technical detail at how these cost savings are achieved.

Reducing Porting Costs

Airplay System uses a combination of a single game binary (i.e., the game code binary is identical on all target platforms), and a thin C API which abstracts all relevant core OS services. This API behaves identically across all target platforms, thus reducing device and OS fragmentation. The game developer is concerned with only one development environment, and does not need to acquire any knowledge of, or even install, the target platform SDK.

The abstraction layer is kept as thin as possible, containing only truly OS-dependent functionality, with all utility functions and other OS-independent functionality implemented in higher-level libraries that are linked into the game binary (if that functionality is needed). The result is that a single game binary typically contains 90 % of all code that is executed, and this is identical on all platforms.

Airplay System takes great care to ensure that the remaining 10 % of code behaves identically, and the abstraction layer can be configured (by either Ideaworks3D or a licensee) for a particular device via a text-based device profile to activate particular functionality or workarounds for handset issues.

Airplay System uses an OS-independent, but CPU-dependent, binary format. This ensures that all code is loaded and executed as pure native instructions, with no virtual machine (VM), ahead-of-time (AOT) or just-in-time (JIT) compilation, all of which degrade performance and increase load times. The vast majority of addressable native operating systems today run on ARM processors; therefore the 'single game binary' is usually a native ARM binary. CPU architectures other than ARM and x86 can be supported if needed.

The game binary is native code,[1] as generated by the best-in-class optimizing compiler for the target architecture (RVCT 3.0 on ARM), thereby maximizing performance. The GCC compiler is also supported. Airplay System supports all compiler optimizations, which can be applied on a global basis or by the use of compiler directives (such as in-lining functions) and/or pragmas to provide fine-grained control. Similarly, Airplay System can generate code for the ARM, Thumb or Thumb-2 instruction sets in order to trade off size versus performance, as appropriate.

Reducing Code Development Costs

Developing for multiple native operating systems is currently a painful process. For each platform the programmer typically has to:

- obtain and install the target OS SDK for each flavor of each OS

- learn the coding idioms and core APIs of each OS

- work within restricted C++ environments, for example, with limited support for standard C++ libraries such as the STL

- use a particular IDE for a particular OS

- test code on the desktop within proprietary OS simulators. These are often slow to boot up, and are still running x86 code, so they are not able to detect issues that may only be evident when compiling for ARM.

- test code on hardware, possibly without any form of on-device debugging, which requires writing trace statements to text files, or infoprint

---

[1] The specification for the game binary file format is included within the Airplay SDK documentation.

statements. (Note that Symbian OS v9 does have good support for on-device debugging and code profiling in the Carbide.c++ IDE.)

This can add up to an extremely cumbersome, time-consuming and costly exercise when working on multiple platforms.

Airplay System removes every one of these restrictions. It abstracts the technical details of the native operating systems, and even details of those operating systems' proprietary SDKs because the developer does not need to install any OS-specific SDKs in order to develop games for those platforms. Airplay System also provides full C and C++ standard libraries, and a full STL implementation.

Because one of the supported native operating systems is Windows, the game developer can do 95 % of their game development purely within a familiar desktop environment. An x86 build of the game can be debugged within the standard desktop IDE debugger. Currently the recommended IDE is Visual C++ (VC6, VC. NET 2003 and VC 2005 are all supported), because this is widely considered to be the most advanced and efficient development environment available. It has full support for both code and data breakpoints, edit-and-continue, and fast compilation, and most developers are already very familiar with it.

The Windows implementation of the solution is designed to model as closely as possible all aspects of the target devices, and to raise issues within the IDE before hardware testing takes place. For example, screen size, available heap memory, available stack memory, portrait or landscape display, CPU/GPU load balancing of graphics pipeline and many more parameters can all be configured and respected within the Windows implementation.

However, there are certain use cases that oblige a developer to test the game within the simulator provided by the native target platform SDK (for example, the BREW simulators), because it may be required by certain platforms for their certification process (for example, TRUE BREW compatibility testing). For such cases, the Airplay System solution seamlessly integrates with the simulator so that a developer can very easily test and debug their application within the development environment.

Reducing Code Debugging

Airplay System includes a real-time ARM emulator that allows execution of the single ARM binary within an ARM debugger (which is also included). This approach uniquely provides the ability to test and debug the exact code that will be run on a device within a desktop environment, and results in a huge increase in productivity, compared to using on-device debugging.

The real-time ARM emulator is fully integrated with Visual C++; if building an ARM Debug configuration, the ARM emulator is launched instead of the Microsoft x86 debugger. Both software rendering and OpenGL ES targets can be debugged at source-level within the ARM emulator.

Performance of the ARM emulator is roughly equivalent to a 150 MHz ARM 9 target device with no hardware graphics acceleration.

Reducing Art Development Costs

Platform fragmentation persists not only for the native OS but also for the graphics capabilities of the device. Addressable devices typically range from ARM9 100 MHz, all the way up to ARM11 400 MHz+ with GPU.

Hardware acceleration can generally be addressed through OpenGL ES drivers; however, OpenGL ES is not a good choice for pure software rendering, and typically developers are required to develop their own software renderer if they want to achieve any kind of interactivity on low-end devices.

Developing a high-performance software renderer is a costly exercise. Even then, the developer must write two sets of render code; one to address their software renderer, and one to address OpenGL ES (the future need to address other graphics APIs such as D3D Mobile will only makes this situation worse).

Airplay Studio solves the graphics fragmentation problem by providing a complete end-to-end tools and runtime art pipeline, which seamlessly addresses low-end pure software rendering devices, high-end hardware-accelerated devices, and all capabilities in between. The runtime provides a graphics API abstraction layer that seamlessly supports software rendering and any other graphics APIs the device may provide. It includes an optimized software renderer (over six years in development) that is widely considered to be the best-performing software renderer in mobile games development today.

The term 'scalable content pipeline' sums up the following features and efficiencies provided by Airplay Studio:

- the ability to export models, animations and materials from 3DS Max or Maya

- software rendering and OpenGL ES 1.1 from a single set of art assets

- software rendering and OpenGL ES 1.1 from a single set of render code

- the ability to down-scale high-end art assets gracefully as required for software rendering; for example, automatically palletizing textures, automatically merging triangles into n-gons for software rendering, automatically tri-stripped for OpenGL ES

- load-balancing between CPU and GPU, according to the maturity of the GPU and drivers.

The final point may benefit from some explanation. At Ideaworks3D, we have been working for several years with prototype and shipping versions of first-generation hardware-accelerated mobile devices. We quickly learned that the OpenGL ES drivers were a critical component in the equation when trying to maximize graphics performance. Just because a GPU driver is compliant with the OpenGL ES specification, it doesn't mean that all the API calls are actually accelerated by the GPU – many will be implemented purely in software. Often these implementations are so slow that the developer, knowing the specifics of their API usage, could do a better job themselves. We found that, to ensure maximum performance across all flavors of devices, there are three stages of the pipeline which need to be switchable between the OpenGL ES drivers and an optimized software solution: transform, lighting, and rasterization. Airplay Studio uniquely provides this flexibility, therefore providing optimal graphics performance across all devices from a single game binary.

Reducing the Cost of Deploying to New Handsets

Airplay splits the production and post-production process, by having separate development and deployment environments, ensuring that game developers and deployment producers can concentrate on their specific tasks alone, resulting in much higher productivity. Airplay is designed to allow additional platforms and handsets to be added, and to allow games that have already been completed to be deployed to these new platforms or handsets, without needing to rebuild the game.

Support for additional platforms is added by the Ideaworks3D team responsible for implementing the platform abstraction APIs. These engineers are experts in the wide range of issues related to working with new platforms, and with all the subtleties involved in ensuring compliant behavior.

Once the additional platform support has been added, an updated version of the solution is distributed with the new deployment target. This deployment target can be accessed via both the development and deployment environments, and it is not necessary to rebuild any projects in order to deploy to the new platform. Therefore, games that have already been completed, and verified to correctly use the platform abstraction APIs, can trivially be packaged for the new platform in the deployment environment, with no input from the game development team, and thus with negligible deployment cost.

The solution offers a threefold approach to adding support for additional handsets:

- each platform implementation is designed and developed to be robust – potential device-specific issues are anticipated and as many parameters as feasible are obtained from the device. This results in a high level of compatibility without any changes needed

- in order to facilitate working around device issues and limitations without requiring a code rebuild, the solution offers a data-driven approach whereby device profiles can be specified (in a text file) to set parameters to correct common behavior faults

- handset manufacturers and carriers inevitably find new and exciting ways to break the platform implementations on their devices, in which case, it may be necessary to make code changes to correct for these. In this case, the implementation is modified by the specialist developer for that platform within Ideaworks3D, and the workaround is applied as generically as possible and either enabled automatically (if it is possible to detect the fault within code) or via a configuration option within the device profile.

Ideaworks3D maintains a formal process for 'certifying' handsets being supported by the solution. This process involves running a large set of unit tests, followed by system testing our previously released titles. Any required device profile changes or workarounds can then be made and tested. Once this is complete, an updated SDK is released with official support for the handset. In practice however, for a platform that has a relatively mature (>5 devices certified) implementation, the handsets usually can be certified without modification, or, at the least, require data-driven rather than code changes to Airplay System. Given that the device profile script can be edited by users of the SDK, in these cases, deployment to new handsets can take place independently of Ideaworks3D.

Figure A.1 outlines the processes for adding support for new devices and platforms to the solution.

Note: OpenKODE

Ideaworks3D has been a very active member of the Khronos working group defining the OpenKODE API, and Airplay will fully support OpenKODE when ratified. The OpenKODE group has recognized that Ideaworks3D has a vast amount of experience both in designing and implementing platform abstraction APIs and in building mobile games

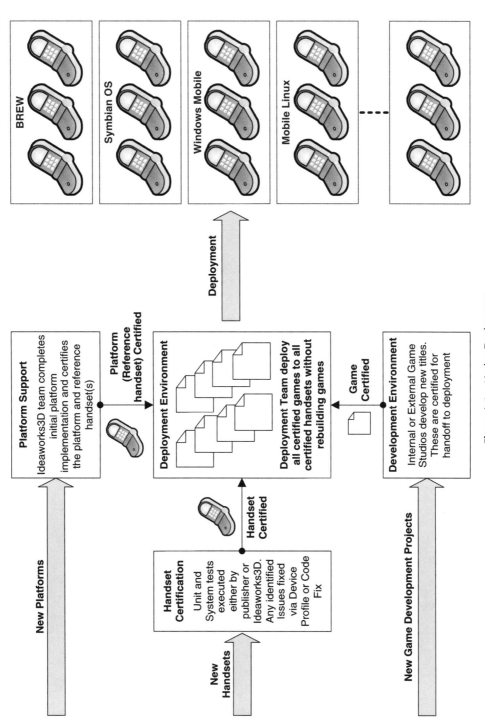

Figure A.1 Airplay Deployment

using such APIs. Many key issues have been raised by Ideaworks3D and embraced by the OpenKODE specification team. Therefore, games developed for OpenKODE and those developed using other Khronos APIs are seamlessly supported within the solution and can take advantage of all the features of both the development and deployment environments.

Index